VOLUME 11

VICKERS
VISCOUNT

By Robin MacRae Dunn

specialtypress
PUBLISHERS AND WHOLESALERS

Copyright © 2003 Robin MacRae Dunn

Published by
Specialty Press Publishers and Wholesalers
39966 Grand Avenue
North Branch, MN 55056
United States of America
(800) 895-4585 or (651) 277-1400
http://www.specialtypress.com

Distributed in the UK and Europe by
Midland Publishing
4 Watling Drive
Hinckley LE10 3EY, England
Tel: 01455 233 747 Fax: 01455 233 737
http://www.midlandcountiessuperstore.com

ISBN 1-58007-065-5

All rights reserved. No part of this book may be reproduced or transmitted in any form or by any means, electronic or mechanical including photocopying, recording, or by any information storage and retrieval system, without permission from the Publisher in writing.

Material contained in this book is intended for historical and entertainment value only, and is not to be construed as usable for aircraft or component restoration, maintenance, or use.

Printed in China

Title Page: *The first V.748D for Central African Airways, VP-YNA (c/n 98), delivered on 1 May 1956. The aircraft possessed high activity de Havilland propellers to assist takeoff performance at hot-and-high airports, and the stouter engine nacelles of the higher-powered Rolls-Royce Dart Mk. 510. This aircraft spent its entire working life with CAA and its successors until retirement by Air Zimbabwe in 1986. (Author's Collection)*

Front Cover: *British Midland's V.814 G-AWXI (c/n 339) is pictured in 1969, during engine-starting. Two BAC One-Elevens share the Gatwick ramp; developed mainly by former Vickers teams, the twinjet was promoted as the "Jet Successor to the Viscount." (J. Roger Bentley Collection)*

Back Cover (Left Top): *The world's first turbine-powered scheduled air service: on 29 July 1950, the earliest Viscount prototype prepares to leave London for Paris and the history books. (ATPH Transport Photos)*

Back Cover (Right Top): *The Viscount 700D externally-mounted "slipper tank" system, which augmented fuel capacity by 290 Imp gal. (Central African Airways via Author)*

Back Cover (Right Lower): *The driving force: the starboard Rolls-Royce Dart 525s of a Viscount 816 of Merpati Nusantara Airlines, PK-RVS (c/n 433). More than any other factor, the world's first turboprop aero-engine was vital to the Viscount's success, and powered many other types besides. The oil cooler intakes are conspicuous atop each nacelle immediately aft of the propeller. (Chris English)*

TABLE OF CONTENTS

VICKERS VISCOUNT

Introduction 4
A Word from the Author

Chapter 1   **Viscount Inspiration** 7
A Turbine Revolution

Chapter 2   **Viscount Airborne** 15
Development and Certification

Chapter 3   **Viscount Assembly** 25
The Viscount 700 Described

Chapter 4   **Building on Success** 39
Developing The Viscount

Chapter 5   **Life With The Viscount** 51
The Day by Day Story

Color Section   **A Lively Career** 65
50 Viscount Years

Chapter 6   **Viscount at Work** 73
Worldwide Service; Worldwide Success

Appendix A   **Viscount Variants** 98
The Vickers Type Numbering Scheme

Appendix B   **Specifications** 101
Principal Viscount and Rolls-Royce Dart Data

Appendix C   **Model Kit Guide** 102
By Richard Marmo

Appendix D   **Significant Dates** 104
Key Dates in the History of the Vickers Viscount

VICKERS VISCOUNT

INTRODUCTION

A WORD FROM THE AUTHOR

"... a machine so patently superior to its rivals that it has set in its field wholly new standards of comfort, performance and ... reliability."
– *Flight,* 14 October 1955

When the first Vickers Viscount lifted off from the grass runway at the company's airfield at Wisley, near London on a damp July morning in 1948, it climbed away not only on its own maiden flight but into aviation history. That modest takeoff, watched by a mustering of cautiously eager Vickers employees, took place with none of the extravagance that one day would be *de rigueur* for a new airliner debut. What the occasion lacked in pretentiousness, however, it possessed a thousand-fold in significance. Those fortunate spectators witnessed not only the first flight of a brand new aircraft, but also that of the world's first turbine-powered airliner. Two years later, the same aircraft would carry the first-ever fare-paying passengers to travel on a scheduled, turbine-powered air service. The event would soon be outshone by the spectacular service entry of the pure jet de Havilland Comet. However, it was the more humble Viscount, propelled by the less dramatic but equally elegant turboprop, that inscribed turbine power into the world airline story.

For Vickers, being first was a glorious and satisfying accomplishment. By the time the Viscount rooted itself in the imaginations of its authors, however, transport aircraft design had already ceased to be a romantic matter of climbing mountains simply because they were there. Experiment remained a necessary element – but the challenge was to create something *useful.* The Viscount came into being to do a job, and Vickers never lost sight of that fact. A masterpiece of straightforward design, it was a turboprop because being a turboprop made economic sense. When asked to explain his airline's remarkable commitment to a large fleet of the novel, non-American airliners, J. H. Carmichael, president of Capital Airlines, simply replied, "We are cold-blooded businessmen." It is little wonder that a mere decade from the prototype's first flight, the Viscount had become not only Britain's most successful airliner ever, but also the world's most widely-used turbine-powered air transport. It would remain so until the pure-jet 1960s.

Yet the Viscount's productivity was also indivisible from its human appeal: the air-travelling public loved it. A smooth journey in which it was possible to chat normally to fellow passengers was a miracle. Those giant windows were a beautifully personal touch, adding spaciousness and making passengers actually eager for the journey. Glamorous but never flashy, the Viscount was an airliner that looked good both in the air and on the balance-sheet – the perfect blend of function and form.

Completing this volume would have been impossible without the kind assistance and support of numerous people and concerns, to all of whom I am greatly indebted: Ron Davies, John and Karen Dunn, Kev Darling at BBA, Capt. Richard Edwards, Peggy Frank, David Gill, Megan Graff, Phyllis and Ray Grumbine, Christine Hardy, Bob Jenner Hobbs and the Association of Transport Photographers and Historians, Antti Hyvärinen, Roger Jackson, Jeff Kohen, Stan Kolodzie, Peter Offerman of KLM Corporate Communications, Leith Paxton, Sveinn

First of the line: an early photograph of the sole Viscount 630, G-AHRF, wearing Vickers-Armstrongs insignia. (BBA Collection)

The Viscount offered numerous smaller operators an opportunity to acquire modern, versatile equipment. Icelandair's two 53-seat Viscount 759Ds – the airline's first pressurized aircraft – flew European and domestic schedules and, under charter to the Danish authorities, often visited Greenland. TF-ISN (c/n 140) was named "Gullfaxi." (Icelandair)

Sæmundsson (Icelandair, 1957-90), Nicky Scherrer, Bob Turner, John Austin-Williams of the South African Airways Museum Society, Air-Britain, and The Washington Airline Society. Carl Ford, Paul J. Hooper, and the personnel at the Brooklands Museum, particularly Mike Goodall and the Curator of Aviation, Julian Temple, were all especially generous. Dave Arnold and Steve Hendrickson supplied much guidance and faith, and I offer them and their colleagues at Specialty Press, especially Dennis R. Jenkins, my deep gratitude. Three people above all whose loyal help shifted the project from an impossibility to a fact are Chris Sterling, Roger Bentley and my wife, Cathi MacRae, whose patience and practical advice made all the difference in the world.

VICKERS-ARMSTRONGS: THE VISCOUNT PEDIGREE

The traditions that eventually brought the world its first turbo-prop airliner lie rooted far deeper than the aviation era. To feed the appetite of Victorian Britain's prodigious military machine, the firms of Armstrong and Vickers grew from inventive producers of engineering equipment and steel castings into vast weapons suppliers, manufacturing everything from bullets to battleships.

In 1911, Vickers established an aviation division, and by World War I was contributing aircraft such as the Gunbus. The Vimy bomber was a heavy, twin-engined biplane. The Vimy found fame in 1919, being flown by Alcock and Brown from Newfoundland to Ireland on the first non-stop, trans-Atlantic air crossing. A few months later, Ross and Keith Smith piloted a Vimy on the first-ever flight from the UK to Australia. The subsequent Vimy Commercial led to a series of large designs, including the Vernon, the company's first true passenger design, and the Virginia, Britain's principal front-line, heavy bomber of the inter-war years. The hefty Victoria was capable of carrying as many as 22 troops. With the Vias-

(Continued on Next Page)

(Continued from Previous Page)

tra transport of 1929, Vickers became one of the first British aircraft firms to explore the benefits of all-metal construction.

In 1927, the foundation of Vickers (Aviation) Ltd. brought autonomy to the aeronautical interests of the newly merged Vickers-Armstrongs conglomerate. Vickers soon acquired Supermarine, renowned for its brilliant seaplane designs for the famous Schneider Trophy contests. In 1929, a Vickers-Supermarine S.6B – the first aircraft to exceed 400 mph – won the Trophy outright. In 1938, corporate reorganization saw the aircraft division re-assuming the Vickers-Armstrongs name.

World War II found Vickers-Armstrongs involved at almost every level. Amazingly, the company was responsible for building 225 warships, including aircraft carriers and submarines, thousands of machine guns, most of the UK's field artillery, and a huge range of tanks and other tracked vehicles. They also built over 33,000 aircraft, of which, two out of every three were versions of the incomparable Spitfire. The Wellington gave vital service as a mainstay of the RAF's (Royal Air Force) medium bomber force in the first half of the war.

Although the aviation arm of Vickers-Armstrongs would provide the UK with further airborne might in the form of the Valiant nuclear bomber, its strategy had already re-focused on commercial flight, engendering one of the most popular and influential of all airliners – the Viscount. Although the Vickers name would disappear in the 1960s in the amalgamation that created the British Aircraft Corporation, the company's expertise and energy would long remain embedded in the UK's aviation industry.

Among the earliest true airliners, the Vickers Viastra of 1930 reflected the collaboration between Vickers and Michel Wibault, the French pioneer of all-metal aircraft construction. Produced in single-, twin- or three-engined versions, it saw commercial service in Australia. G-ACCC – built as the personal transport of the Prince of Wales, and thus possibly the first executive airliner – is seen after conversion for experimental radio trials. Note the early form of leading-edge slats. (The A. J. Jackson Collection)

VISCOUNT INSPIRATION
A TURBINE REVOLUTION

At the close of World War II, the British aero-engine industry led the world in turbojet development, and stood alone in researching the turboprop. While the seminal work of Frank Whittle treated the gas turbine as a means of eliminating the propeller and focused on the pure jet, most thinking on the subject sprang from the long-time search for an alternative to the piston engine as a method of driving a conventional airscrew. As early as 1929, A. A. Griffith of the Royal Aircraft Establishment had advocated marrying the simplicity of the gas turbine to the effectiveness and responsiveness of the propeller. Rather than exiting exclusively as a propulsive stream, the jet thrust would drive a turbine and thereby turn a propeller, resulting in a powerplant that retained the advantages of the piston engine, but was far less complex.

Rex Pierson, chief designer of Vickers-Armstrongs' Aircraft Division, quickly appreciated the transformative impact such a powerplant might have on air transport economics. More efficient than the pure jet at lower altitudes and modest velocity, and at home on shorter runways, the propeller-turbine would be neatly suited to European air services. From 1943 onward, Pierson's team explored turboprop derivations of the four-engined Vickers Windsor, a planned long-range bomber that, although cancelled, still offered possibilities as a transport.

Pierson was not alone in thinking ahead. In a striking gesture of faith in a favourable outcome to the war, in late 1942 the UK government set up a committee chaired by Lord Brabazon to define commercial aviation needs for a postwar Britain. This committee identified four categories of necessary transport aircraft. Tasked with fleshing out these requirements, a second Brabazon Committee added a fifth classification and, between August 1943 and November 1945, progressively issued detailed specifications. These specifications called for a heavy, trans-Atlantic transport (built as the gigantic, abortive Bristol Brabazon), a medium/long-range airliner for Britain's colonial routes (ultimately answered by the Bristol Britannia), a jet-powered trans-Atlantic mailplane (which evolved into the de Havilland Comet) and a piston-powered feeder-liner (prompting the highly successful de Havilland Dove). Additionally, the committee's second classification, or Type 2, summarized an airliner for European services of moderate stage-lengths. Initially, this definition generated the piston-engined Airspeed Ambassador, but would find its lasting answer in the Vickers Viscount.

Despite such planning, war's end found the UK with barely any indigenous transport aircraft, yet facing an urgent need for civil air capability. Most large British transports were military aircraft converted merely by removing armament and installing whatever seating might be made to fit. Soon, specialized civil variants arrived, incorporating revisions such as cabin windows and re-fashioned interiors. A longer-term answer, stimulated by a subsidiary recommendation of the first Brabazon Committee, was to develop interim civil types by designing from scratch what was

By the time of this 1950 photograph, the Viscount had survived many setbacks to emerge as a highly promising new air transport. A half-century of service lay ahead. (BBA Collection)

VICKERS VISCOUNT

Large aircraft design constantly exercised the inventiveness of Rex Pierson, Vickers' chief designer. His Windsor bomber, powered by four Rolls-Royce Merlins, arrived too late to see World War II service, and only three flew. However, studies for transport derivatives – including some with turboprops – set Vickers on the road to the Viscount. Note the main undercarriage units – one per nacelle. (Brooklands Museum)

necessary and combining it with major elements of a proven military type. Thus, Avro produced the utilitarian York by attaching a new fuselage to the wings, engines and empennage of the Lancaster. Vickers, knowing any wholly new type could not be ready for several years, filled the gap with an expedited airliner of its own.

The Vickers Viking

The resulting twin-engined, tail-wheeled Vickers VC1 (Vickers Commercial 1) was conceived to supplement, and partly to replace, the hard-working Douglas DC-3. Soon named Viking, the VC1 mated the wings and undercarriage of the Vickers Wellington bomber, plus the tail surfaces of the related Warwick, to a new, all-metal, stressed-skin fuselage. It also employed a civilianized version of Wellington's Bristol Hercules sleeve-valve piston engines.

The Viking first flew in June 1945 and entered service with the newly organized British European Airways (BEA) on 1 September 1946. This initial model accommodated 21 passengers, and soon gave way to an upgraded version of similar capacity but with stressed-skin mainplanes replacing the fabric-covered geodetic outer wing panels. A third variant featured a slightly lengthened fuselage seating 27, later increased by BEA to 36. The Valetta was a militarized edition, while a nose-wheeled development named the Varsity served the Royal Air Force (RAF) as an advanced aircrew trainer. Presaging similar trials with the Viscount, one Viking even became the world's first jet transport – albeit purely as an experiment – re-engined with two Rolls-Royce Nene turbojets.

Despite being an extemporized design lacking a pressurized cabin, in most respects the Viking was up to date and became a great success. In front-line operation for only a few years, Vickers' rugged airliner nevertheless worked unremittingly for second-tier carriers until as late as the 1970s. Production totalled 163, with nearly one third for export;

Vickers also built well over 400 military derivatives. With the Viking, the company gained essential experience in modern airliner manufacturing and established a crucial customer base. In particular, Vickers founded a relationship with BEA that would prove indispensable as its designers strove to create the world's first propeller-turbine airliner.

The Beginnings of the Viscount: The Vickers VC2

The Viscount originated less in a specific Brabazon Committee recommendation than in a happy confluence of the committee's visionary sponsorship of the gas turbine – both the pure jet and its inspired derivative, the turboprop – and the insights and energy of Vickers-Armstrongs' own design department. In December 1944, Rex Pierson attended a meeting of the second Brabazon Committee. There, discussion on how the forthcoming Viking might match the short-haul Type 2 specification extended to Pierson's ideas for a Viking successor, including a pure jet adaptation. Subsequently, both the committee and Pierson concluded that a propeller-turbine airliner held superior promise.

Soon, the "new Viking" studies merged with those for a turboprop Windsor. Under the general Vickers type number 453, the latter had produced many variations, notably the V.601 driven by four Rolls-Royce Clyde turboprops producing 3,040 hp each. (Remarkably, this Windsor model would have been capable of over 400 mph at 28,000 ft.) By March 1945, all V.453 plans had coalesced into a fresh creation called the VC2, distinguished by a new, Viking-influenced wing of shorter span. Outlines submitted to the Ministry of Aircraft Production (MAP) described pressurized and non-pres-

surized versions of a 24-passenger turboprop airliner of 34,200 lb., plus an unpressurized 27-seater, each with a no-reserve, still-air range of 1,000 miles.

Lord Brabazon at once urged the MAP to consider the VC2 a Type 2 contender. That month, the committee issued a rewritten Type 2 requirement, acknowledging the potential of turboprop power. The original recommendation became the Type 2A, while a new Type 2B specification outlined a pressurized, 24-seat, four-engined, propeller-turbine aircraft, capable of flying 1,040 miles with a payload of 7,500 lb. The MAP encouraged Vickers to fine-tune the VC2 to this description, intimating that an official order for four prototypes would follow.

A 24-seater suggested simply a Viking replacement. Indeed, in the first official VC2 response from Vickers, the design's tubbiness, tail layout and mid-wing all betrayed Viking genes. Very different, however, were a fuselage of "double-bubble" cross-section, a slightly swept-back wing and a tricycle undercarriage. Gross weight was to be 24,500 lb., still-air range 1,040 miles and cruising speed 297 mph at 20,000 ft. Interestingly, even at this time, the proposed turboprop powerplants were four centrifugal-flow 1,030 ehp Rolls-Royce Darts, although Napier Naiad, Fedden Cotswold and Armstrong Siddeley Mamba engines (all axial-flow designs) were active alternatives. Also already present were the cockpit shape and large, elliptical cabin windows that would be so characteristic of the eventual Viscount appearance.

On Pierson's promotion to chief engineer in September 1945, George Edwards – hitherto Vickers' experimental manager – assumed authority over all design effort. Immediately,

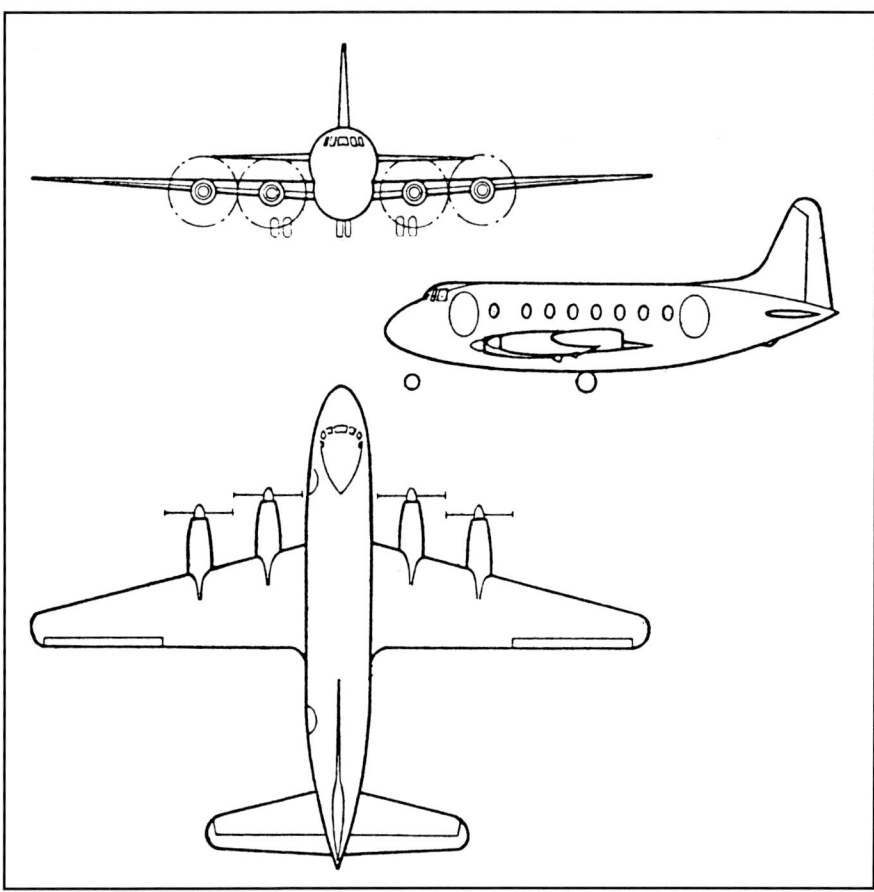

This diagram from June 1945 depicts the first firm VC2 layout: note the wing leading edge sweepback and the double lobe fuselage. Both would disappear three months later when George Edwards took over as chief designer and decided on the familiar Viscount pattern of circular cabin cross-section and straight, evenly tapered wing. (Vickers-Armstrongs via Author)

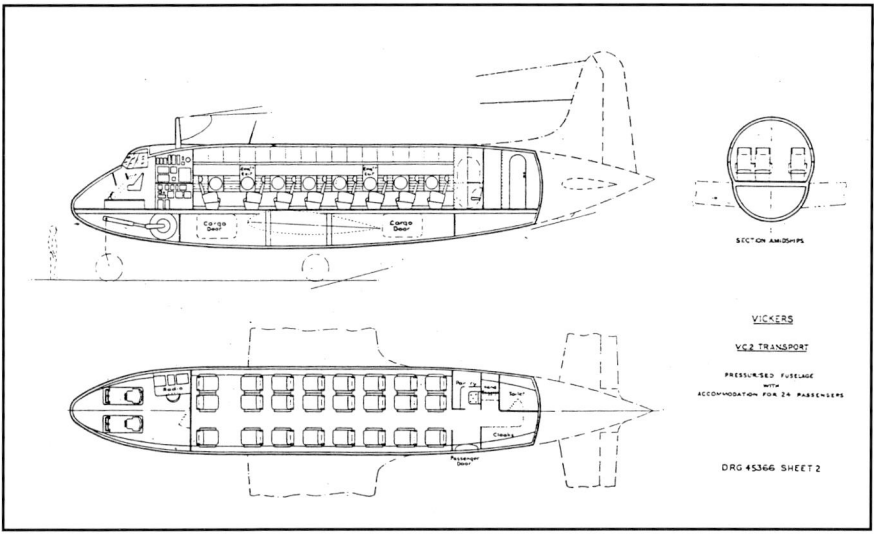

A cutaway view of the VC2 as proposed in mid 1945, showing the underfloor hold that gave rise to the double-bubble fuselage; note also the three-abreast layout and the tail design reminiscent of the Viking. (Vickers-Armstrongs via Author)

VICKERS VISCOUNT

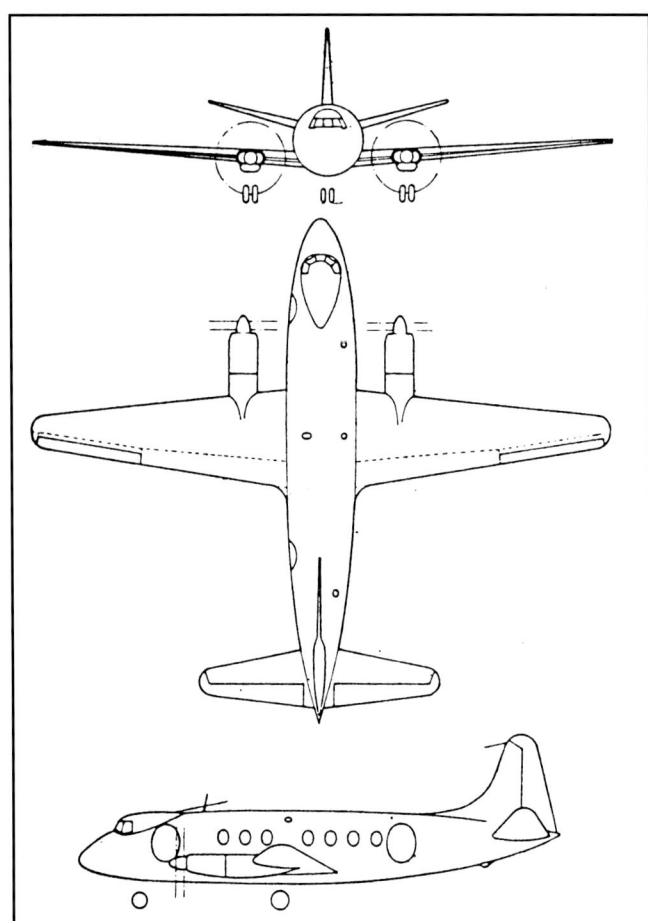 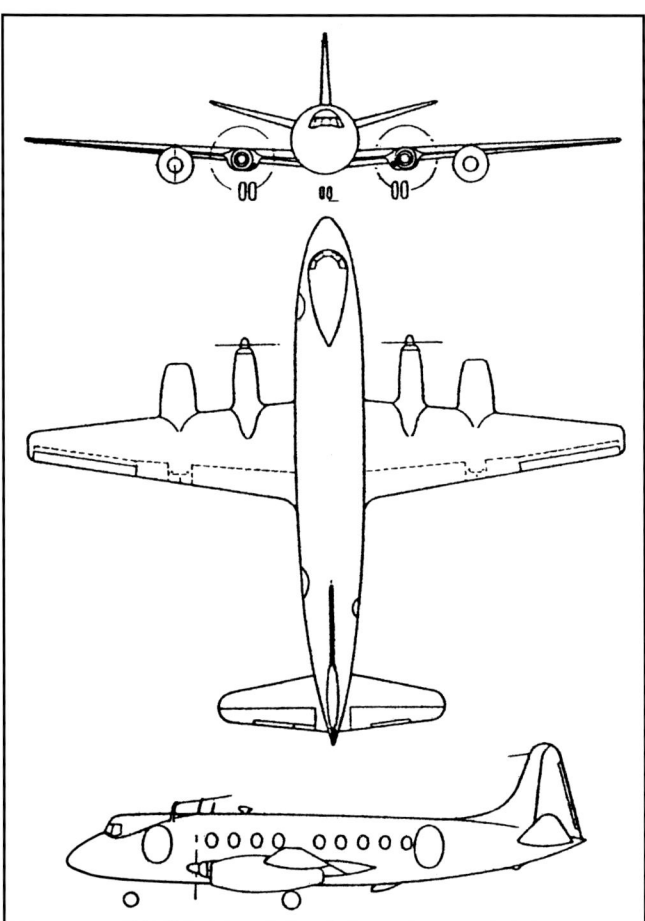

One of the supplementary VC2 studies in 1946 (at left) was for this version powered by two Napier Double Naiad turboprops. Another VC2 proposal (at right) suggested powering the Type 609, eccentrically, with inboard Rolls-Royce Darts and pure jets – either Derwents or Nenes – in the outer positions. In this diagram, the VC2 has a longer fuselage with added windows. (Vickers-Armstrongs via Author)

the man whose conviction was to push the nascent airliner towards triumph set about a review of the VC2 program, triggering several important design decisions. The most significant concerned the combined question of pressurization and fuselage shape – about which Vickers, thanks to experience with pressurized Wellington projects, possessed much wisdom. To make economic sense, turboprop engines needed to operate at 20,000 ft. or above, rendering cabin pressurization inescapable. Doubts remained over the resilience of the VC2 fuselage shape, given the proposed pressure differential of 6.5 psi. Sacrificing the underfloor holds that had dictated the double-bubble configuration, Edwards introduced a stronger fuselage of circular cross-section. At an external diameter of 10 feet, this incidentally established the final Viscount hull-width. Other changes included the birth of the distinctive Viscount dihedral tailplane, while the all stressed-skin wing – derived, with its single main-spar, from that of the Viking, and dating back to the Wellington – acquired a straight taper. The inward-retracting main undercarriage yielded to a forward-retracting arrangement, thus permitting emergency extension using gravity and slipstream-effect, with stowage in the inboard engine nacelles. Meanwhile, disappointing tests overshadowed the Dart engine, pushing the comparable Armstrong Siddeley Mamba and slightly more powerful Napier Naiad to the fore. Judiciously, Vickers worked with all three firms to evolve a nacelle capable of housing whichever powerplant was eventually chosen.

By the end of 1945, the VC2 had become a 35,500 lb. airliner powered by four turboprops of 1,000 shp each, accommodating 24-28 passengers seated in pairs, with all baggage holds on the main cabin level. Wingspan was 88 ft., with a gross wing area of 860 sq ft., and fuselage length 65 ft. 6 in. With full payload, the no-allowance range remained 1,040 miles, with a maximum of 1,380 miles, cruising at 294 mph and 20,000 ft. On 9 March 1946, the ministry of Supply (MoS, successor to the MAP) placed a contract for two VC2 prototypes based on this outline. However, the official Specification 8/46, issued a few weeks later, requested – under pressure from BEA – that the new transport be capable of carrying 32 passengers, compelling Vickers to extend the fuselage to 74 ft. 6 in. and the wingspan to 89 ft. Required freight capacity was 275 cu ft., with seating readily removable for freight operations. Additionally, citing superior test results, the MoS insisted on the Mamba powerplant. These stipulations resulted in a revised weight of 38,170 lb. Stringent performance requirements included a ceiling of 30,000 ft., a stalling speed of 81 mph in landing configuration, and the ability to take off in under 3,600 ft. on three engines. Also specified were a cabin pressure differential of 6.5 psi, emergency oxygen for the flight-crew, and – interestingly – a maximum cabin noise level of 60 dB. This revised VC2 acquired a fresh type number, becoming the V.609. As if to stress the project's permanence, the manufacturer also endowed its plain "Vickers Commercial 2" with a name: Viceroy.

The interim VC1 Viking gave Vickers invaluable experience in the manufacture and marketing of modern transports, and provided the company with much of the design basis of the Viscount. The Viking sold well and served widely, including in India and Africa. (Author's Collection)

COMPETITION:
ARMSTRONG WHITWORTH APOLLO

To many, VC2 progress during 1945 seemed sluggish. Late in the year, an itchy Brabazon Committee pressed the MoS to guard against failure by inviting four other manufacturers – Handley Page, Short Bros., Armstrong Whitworth, and Blackburn – to tender. Early in 1946, the MoS approved Armstrong Whitworth's submission for their turboprop AW55, powered by four axial-flow Armstrong Siddeley Mambas and seating 26-31. Soon named Apollo (after fleeting spells as the Achilles, then Avon), the AW55 featured a fuselage measuring 64 ft. (eventually increased to 71 ft. 6 in.) and a wingspan of 92 ft. At 45,000 lb., it was rather heavier than the Viceroy. Two flying prototypes and a static-test fuselage were constructed. The Apollo took to the air in April 1949, but by that time the Vickers aircraft – and the Airspeed Ambassador – had clearly surpassed

The first Viscount – the Type 630 – taking shape. The inner wing is in place, but the cockpit dome has not yet been fitted. Note the formers in the rear door aperture. (Brooklands Museum)

In the spring of 1948, even though the Viscount suddenly seemed to have little future, Vickers pressed on with completing the first airframe. In this view, the nacelles have been installed. (Brooklands Museum)

The Armstrong Whitworth Apollo offered a brief challenge to the Viscount for the Brabazon Type 2A turboprop requirement. G-AIYN (c/n 3137), first of the two built, is seen after receiving a larger fin and tailplane to counteract stability troubles. Just discernible is that four-bladed propellers have been fitted to the inboard engines. Apollo development ceased in 1952. (The A. J. Jackson Collection)

it. Moreover, tribulations with longitudinal and directional stability necessitated a redesign of the vertical stabilizer and tailplane; less easy to correct were frustrations with the output and reliability of the powerplants. Despite ideas for a larger version and brief interest from one or two airlines, fortune failed to smile on the Apollo: for all its elegance, it never entered production, and the two prototypes spent their days in experimental work.

The Viscount Finally Takes Shape

Although other Viceroy variations crossed the drawing-board after the arrival of the official contract in spring 1946 – one featuring two Napier Double Naiad turboprops driving contra-rotating propellers, another even proposing a combination of Dart turboprops inboard and Derwent or Nene pure jets outboard – Vickers remained convinced that the V.609 represented the best path towards a working airliner. However, the company had expected an order for four prototypes. Believing two would be inadequate for the development program, particularly in view of the several powerplant options, Vickers resolved to fund a third on its own. Despite the reduced MoS order, official support was unambiguous, emphasized by a grant of £1.8 million towards development costs.

At the company's experimental workshops at Foxwarren, construction of the prototypes finally commenced in December 1946. The following August, an updated specification revealed a 32-passenger aircraft of 38,650 lb. Under pressure from the MoS, the powerplant remained the Armstrong Siddeley Mamba, whose trial results continued to be more convincing. Officially, Vickers kept an open mind – but the Dart's ability to defy icing, and the intrinsic robustness of its centrifugal compressor compared with the more complex, axial pattern of its rival, left George Edwards in no doubt. He trusted, too, Rolls-Royce's long experience in turbocharger design. In fact, the tide was about to turn against the Mamba: in late summer, 1947, the MoS altered its opinion in light of dramatically improved Dart performance, and requested that the second prototype be built with Rolls-Royce engines. Vickers needed little persuasion, and, meriting a revised type number, the Dart-powered variant became the V.630.

The entire project also underwent another, and final, name change. In mid year, the UK relinquished its colonial control of India, thereby also rendering obsolete the title of Viceroy – a rather grand form of Governor. Diplomatically, Vickers followed suit, and the Viceroy finally became the Viscount.

Enter the Naysayers – and the Airspeed Ambassador

British European Airways became linked with the VC2 in mid 1945. The airline's Chief Project and Development Engineer, R. C. Morgan, worked extensively to integrate the VC2 with BEA's strategy, also becoming personally involved in planning the cockpit and designing the cabin interior. Vickers naturally considered the publicly owned airline a guaranteed customer, and it came as a stunning shock when, in December 1947, the MoS announced a firm order on behalf of BEA for the rival Airspeed Ambassador. Worse, it seemed clear the Ambassador had been picked instead of – not to supplement – the Viscount.

In truth, the rebuff did not come unforewarned. The Viscount had been in competition with the Ambassador all along. Conceived to meet the original Brabazon Type 2 description, the high-winged Ambassador was powered by two Bristol Centaurus piston engines of 2,625 hp each. With seating for 47, its capacity was actually greater than that of the V.630. Furthermore, although BEA's preference for the

Vickers turboprop had seemed secure, during 1947 doubts had surfaced about the prudence of committing to unproven technology. More particularly, the Viscount's operating economics – especially in comparison to the Ambassador, which was already flying – had come into question. European traffic was increasing more rapidly than expected, leaving the V.630 too small for the task. Some also felt that substandard ATC coverage for higher cruising levels would compel any turboprop airliner to operate at inefficient altitudes. A BEA study claimed the present project would make sense only if re-engined with Rolls-Royce Merlin piston engines, and that in order to be viable as a turboprop the Viscount should be enlarged.

Taking the hint, Vickers reacted with the V.652, with two Bristol Hercules piston powerplants, and the stretched, 40-seat, four-turboprop V.653. Despite being predicated on hopes of a more powerful Dart, the latter scored at once. An eager BEA suggested facilitating the V.653 by first fitting two Centaurus piston engines, reverting when possible to four suitably evolved Darts. The airline's government paymasters demurred. The plan offered an airliner too similar to the Ambassador, which in any case would be available sooner. BEA fell into line.

Dejected, Vickers persuaded the MoS to let it complete the first prototype. Partly to recover something from its investment, the MoS took authority over the second prototype, which would now be powered by turbojets and used for research. Vickers stopped work on its self-financed third airframe, which had become the prototype for an increased-range Viscount, the V.640, propelled by four Napier Naiads. In the upheaval, the Armstrong Siddeley Mamba finally dropped out: the first Viscount would now be built as a V.630 with Rolls-Royce Darts.

Rolls-Royce Dart:
The World's First Original Turbo-Propeller Aero-Engine

In 1944, Rolls-Royce, who had already run a centrifugal-flow jet based on Frank Whittle's ideas, turned to the propeller-turbine. The result was the two-shaft, 4,000 shp RB 39 Clyde. First bench-tested in August 1945, this axial-compressor design flew experimentally some years later, but, being technically over-ambitious and lacking an application, it remained undeveloped.

More significant was the RB 50 Trent (unrelated to today's similarly-named turbofan), a Rolls-Royce Derwent II turbojet with a cropped compressor, single-stage reduction gear, driveshaft and five-bladed propeller. Substituted for the Derwents in a twin-engined Gloster Meteor I fighter, the Trent first flew on 20 September 1945. This scantily remembered event, which took place from the Rolls-Royce jet experimental base at Church Broughton, with Gloster chief test pilot Eric Greenwood at the controls, marked the world's first flight by any turboprop-powered aircraft.

Even as the Trent Meteor took off, studies had begun that would lead to the Dart, known initially as the RB 53. Rolls-Royce chose to stay within its areas of experience: unlike the Clyde, the Dart was to be a centrifugal-flow design. Unfortunately, first efforts, aimed at producing a turboprop of 1,000 shp, resulted in a badly overweight engine requiring a

The V. 630 undergoing final assembly at Wisley. The flaps, powerplants and outer wings remain to be installed, although the torque tubes for operating the flaps are in place along the trailing edge. The partly complete rear passenger door can be seen lying in the foreground. At right, assembly has progressed considerably, with the outer wings and flap assemblies now added. Some details, such as the wing-fuselage fillets, remain to be completed, but rollout is near. (Brooklands Museum)

A very early view of the unpainted V.630 showing its G-AHRF registration. The aircraft's dimensions – considerably smaller than those of production models – are especially apparent, notably in the slimness of the engine nacelles. (Brooklands Museum)

On 16 July 1948, G-AHRF takes off from Wisley on its historic flight, flown by company chief test pilot Mutt Summers and assistant chief test pilot Jock Bryce. During the short sortie, the number two engine JPT gauge went unserviceable – the only snag. (Author's Collection)

radical redesign. Moreover, in bench tests it produced barely 60 percent of its intended power. Troubles piled up: impeller cracks, reduction gear failures, seemingly incurable combustion chamber leaks, and even disintegration.

But a prize was in view: the Vickers VC2 had emerged as a potential Dart application. Numerous changes in design, materials, and assemblies – in addition to no small amount of persistence – gradually brought better results. In October 1947, the new powerplant made an encouraging airborne debut, installed in the nose of an Avro Lancaster. Soon, a Wellington re-engined by Vickers became the first aircraft to fly with the Dart as sole power; later in 1948, the Viscount prototype began to add to Dart experience. The air-test program continued for some years: in 1949, a Rolls-Royce-owned Dakota (Douglas C-47) received two Darts; later, two BEA Dakotas were similarly equipped. In an ironic turn, an Ambassador also eventually flew as a Dart test-bed.

Vickers Presses On:
The Viscount Becomes a Reality

George Edwards had no intention of allowing the Viscount to slip away. Progress had been good. The Dart trials were now achieving healthy results, and BEA remained receptive. Work continued at a subdued rate. Eventually, the first completed fuselage was moved to Wisley for final assembly. Dart R.Da.1 Mk 502 engines were installed, with four-bladed Rotol propellers. Early in June 1948, unadorned by any markings, save the registration G-AHRF, the V.630 prototype finally emerged from the hangar. By Friday, 16 July, she was ready to fly. With Vickers chief test pilot Captain J. "Mutt" Summers in command, assisted by G. R. "Jock" Bryce, the Viscount first made several high-speed runs. Finally, the aircraft left the ground and flew out into an overcast morning.

The historic, 20-minute flight was not only the first by any turbine-powered airliner, but also the first by any aircraft designed and built with – rather than adapted for – turboprop engines. Further trial flights followed.

On 2 September, Summers – who had also been first to fly the Spitfire – demonstrated the new aircraft, now in Vickers house livery, in public for the first time. Days later, the Viscount participated in the flying display at the Farnborough Air Show, after which it began an extensive test program, including an early demonstration visit to France. Since it seemed the Viscount might remain purely experimental, trials were now conducted under the auspices of the MoS, who had supplied the funding. Thus, G-AHRF earned the military serial VX211 and RAF insignia in place of its civil identity.

The V.630 would not become the definitive Viscount. Its work would be to validate the airframe and the practicability of turboprop power. Nevertheless, the protracted design process had resulted in a more mature prototype than most, and the V.630 presented amazingly little trouble during the test program. Its pilots liked it enormously; and everyone else marvelled at its quietness and smooth ride. Although not truly revolutionary – other than in terms of its powerplants – the Viscount marked a vast advance and proved itself scrupulously thought-out.

Immediate success in the developmental trials brought a more buoyant mood to Vickers, and spurred progress towards the classic airliner that would one day abundantly repay the company's confidence.

VISCOUNT 2 AIRBORNE

DEVELOPMENT AND CERTIFICATION

Through the dispiriting days of 1947–48 – with the death of the aircraft's author, Rex Pierson, adding solemnity – Vickers stood faithful to the Viscount. The company first sought ways of turning the V.630 into a commercial proposition by replanning the interior to accommodate as many as 43 passengers, although this came at the expense of other attributes, such as the galley. But, even as the prototype undertook its first sorties, progress at Rolls-Royce rendered such improvisation felicitously redundant. A 40 percent increase in the available power of the Dart – to 1,400 shp – injected sudden life into the proposed larger Viscount.

As a starting point, Vickers revisited the V.653, one of the attempts to parry the Ambassador. Against the V.630, this plan featured a 5 ft. greater wingspan (thanks to a new inner section), a fuselage longer by 6 ft. 8 in. (split equally fore and aft of the main spar), a gross weight raised to 45,000 lb., and capacity for 43 passengers. A project design group reworked the V.653 into the V.700, announced in January 1949. Dimensionally identical, but with initial gross weight raised to 48,000 lb. and maximum cruising speed to 318 mph, the V.700 offered seating options for 40, 48, or 53 passengers. Other differences included an entirely new fuel system and nacelles redesigned to house the larger engines. A supportive MoS duly ordered one V.700 prototype, employing funds released by diverting the second Viscount airframe into military research with a separate budget. Vickers began work using components of the partially built but abandoned third prototype, constructing the fuselage and wings separately at plants near Southampton and Swindon.

The V.630 kept up its development flying regime. In December 1948, Vickers offered the press an opportunity to sample the airliner. In print, the amazed passengers rewarded the turboprop with superlatives for its startlingly low noise and vibration levels as well as its rapid takeoff and climb. Vickers also continued to court BEA. Late August 1949 found the aircraft, restored to civilian identity as G-AHRF and with BEA insignia added to its Vickers-Armstrongs livery, arriving for a demonstration at the Corporation's Northolt base. The Viscount's fortunes now turned an exciting corner: BEA and BOAC (speaking for its subsidiary, British West Indian Airways) announced that they would buy the V.700, so long as the aircraft met its stated performance projections.

On 15 September 1949, after 160 flights and 290 hours of flight-time during 14 months of trials, the V.630 prototype received the first Certificate of Airworthiness issued to any turbine-powered transport aircraft. However, the permit was restricted. It did not cover service in tropical or icing conditions, and excluded pressurized operations, since the pressurization system, although installed, remained unproven.

Pressurization trials took place that autumn, during which the Viscount flew to over 30,000 ft. and practiced emergency descents to test the system's integrity. January 1950 found the aircraft based at

Although the flight trials of the V.630 prototype were trouble-free, in autumn 1948 the program looked very unlikely to progress. For a while, the V.630 became more a government-funded research project than an airliner with a future, and accordingly assumed military markings, as seen in this photograph of the aircraft as VX211, flying on two engines. (Author's Collection)

VICKERS VISCOUNT

Mutt Summers guides G-AHRF, flaps retracted, over the grass at Wisley. The inner props cleared the ground by only seven inches. (Charles E. Brown via Author)

Shannon, engaged in tests of the thermal deicing system – the first in any transport aircraft – which functioned flawlessly in temperatures down to – 18° C (0° F). Ice was unable to form at all if the deicing system was activated before entering a risky area. More remarkably, with wings and tailplane copiously ice-covered, the Viscount continued to fly easily with all deicing equipment switched off. Nevertheless, although the engines never hesitated, ice-accretion in the unprotected intake-mouths gave some concern, largely due to the attendant increase in fuel consumption. Other icing uncertainties involved the propellers and windscreen. Production Viscounts received intake lips equipped with electrical heating, electrical overshoes to replace the fluid-type propeller de-icing, and an improved windscreen deicing-fluid pump.

Only tropical trials remained. In the meantime, the first Viscount prototype was about to be joined by a second. Few would set eyes on it. It would be a Viscount with an extraordinary difference and a curious, shadowy life.

A unique photograph of G-AHRF at the SBAC display at Farnborough on 11 September 1948, in the company of the V.618 "Nene Viking" G-AJPH (c/n 207) – the world's first pure jet transport. The similarity of the latter's engine and main undercarriage installation to those of the later jet-powered Viscount is striking. (via Mrs J. V. Rann)

The V.663 Tay Viscount

The single Viscount not powered by Rolls-Royce Darts emerged from the upheavals of early 1948. The MoS chose to retain the orphaned second prototype for experimental work, much of which would involve high-altitude flying better suited to a pure jet. During production, Vickers therefore replaced the planned Darts with two Rolls-Royce Tay R.Ta.1 centrifugal-flow turbojets each of 6,250 lb. thrust. As with the turbojet Viking, these were installed in underwing nacelles presciently similar to those of the Boeing 737-100/200. The main undercarriage consisted of four separate members retracting into the engine pods, one on either side of each jet-pipe, and the cabin was equipped for test instrumentation. Otherwise, the basic V.609/V.630 airframe was unmodified – which now seems unnervingly innocent, in view of later industry experience with metallurgical stresses imposed by high altitude, repeated pressurization, low temperatures, and high velocities. Redesignated V.663, the unique aircraft first took to the air at Wisley on 15 March 1950, piloted by Jock Bryce.

Details of the Tay Viscount's career are blurred by the covert nature of its trade. It never carried its allotted civil registration, G-AHRG, instead wearing the military serial VX217 and RAF roundels on its otherwise unpainted fuselage. In September 1950, it participated in its only public appearance, the flying display at the Farnborough Air Show. It then flew for the MoS, based at Seighford, and assisted for a time in experimental trials with Louis Newmark Ltd. and the Decca Navigator Co. Notably, VX217 flew from Defford for Boulton Paul, assisting development of the powered control system for the Vickers

V.660 Valiant bomber. This experimental work occupied the Tay Viscount for some years, during which it quietly made history by becoming the world's first aircraft to fly with an electrically signalled flight-control system – nowadays known by the term "fly-by-wire."

No performance figures for the only jet-powered Viscount have ever been released, but it is reported to have been capable of speeds as high as 550 mph at 30,000 ft. Contemporary observations suggest that it was an aircraft clearly more animated than other sizeable jet aircraft of its era. In 1958, a hydraulic fire in a wheel bay severed the main spar, irreparably damaging the airframe. With just over 110 hours logged, the pioneering V.663 was scrapped in 1960.

BEA and the World's First Turbine-Powered Air Service

On 20 March 1950, repainted in a new BEA livery, the V.630 prototype began a demonstration tour of eight different European cities, involving 70 flights, 61 airborne hours, and 4,400 miles. The Viscount used only 17,423 Imp. gal. of kerosene; just as remarkably, oil consumption was only 7.9 Imp. Gal. A claimed Viscount feature was an ability to descend and "loiter" on only two engines to save fuel. During the tour, the crew demonstrated the technique several times. The practice was never adopted in service, partly because of worries over compromised performance should an overshoot prove necessary.

During tropical certification trials in June and July at Nairobi and Khartoum, BEA's Captain R. "Dickie" Rymer – the world's first airline pilot to hold a licence endorsed for turbine operations – captained the aircraft in transit and acted as copilot for most

Propelled not by turboprops but by two Rolls-Royce Tay turbojets, the lone V.663 (c/n 2) served as an experimental vehicle involved in largely secret work, and undertook pioneering investigations into powered control systems. VX217 is seen in 1950. (The A. J. Jackson Collection)

An extremely rare glimpse of the Tay-powered Viscount at Seighford on 11 July 1959, the year after fire damage forced its retirement. Note that the scoop air intake has been relocated from the aircraft underside to above the rear entrance. Also clear is the reduction in flap span to accommodate the jet-pipe. (ATPH Transport Photos)

of the hot-and-high test flying. Among other achievements, the Viscount undertook takeoffs from Eastleigh, Nairobi, located at 5,371 ft. above mean sea level (amsl). Although the Viscount was unassisted by water-methanol injection, it completed the takeoffs without any diminution in performance. The Viscount had won its spurs with the completion of these tropical tests. On 27 July 1950, the Air Registration Board (ARB) granted the V.630 a full, unrestricted passenger-carrying Certificate of Airworthiness. On the following day, an unladen G-AHRF flew a London-Paris trial round-trip – an unobtrusive foretaste of excitement to come.

On Saturday, 29 July 1950, at London's Northolt airport, an all-BEA crew welcomed 14 fare-paying passengers on board G-AHRF. Twelve special guests joined them, including Sir Frank Whittle and George Edwards, whose personal energy had done so much to push and pull the Viscount to this auspicious day. Bound for Paris on a standard BEA schedule, with Captain Rymer in command, the V.630 Viscount left the runway at 12:48 to begin the world's first scheduled airline journey under turbine power.

G-AHRF (c/n 1) at Wisley under a threatening sky. The V.630 was fitted with a tail bumper, deleted on later models. (BBA Collection)

In 1950, G-AHRF was repainted in a new white-top BEA scheme before setting off on a European demonstration tour. (BBA Collection)

wholly reliable, but the 1,815 appreciative passengers encountered an utterly unfamiliar style of air travel: one that was miraculously smooth, blissfully kind to the ears, and free of such drudgery as pre-takeoff engine run-ups.

The public relations reward earned by these proving flights provided an ideal – and doubtless planned – background to a long-awaited announcement. On 3 August 1950, BEA confirmed an order for 20 Viscounts. This landmark contract, the first concrete Viscount order, specified the V.701, an enhanced version of the imminent V.700 prototype.

The First Viscount 700 Flies

Days after the conclusion of the operational trials with the V.630, the larger Viscount 700, adorned in BEA insignia and registered G-AMAV, made its debut. The true prototype of an airliner that would become familiar across the world took off for the first time on the afternoon of 28 August 1950, in the hands of Jock Bryce, leaving the Weybridge run-

After 57 minutes, the Viscount touched down at Le Bourget airport, having carved one-third off the usual schedule. The historic journey opened two weeks of service on the London-Paris route, followed by a similar routine between London and Edinburgh. Involving 44 return trips and 127 flying hours, the trials offered useful experience of everyday Viscount operation for BEA flight crews, cabin staff, ground crews, and air traffic controllers. Not only did the Viscount prove itself

George Edwards steered the Viscount towards success – first as chief designer, then as managing director of Vickers-Armstrongs. Without his conviction and leadership, especially when official support dried up, the Viscount might well have withered away. His imprint also was on the Viking, and later on the Vanguard, VC10, BAC One-Eleven, and Concorde, among many others. He received a knighthood in 1957 and retired as chairman of the British Aircraft Corporation in 1977. (Brooklands Museum)

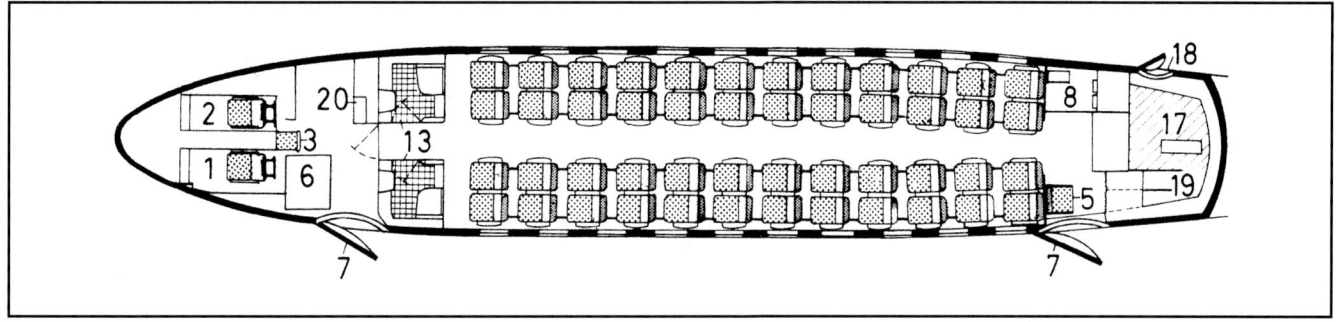

The 48-seat interior of the two-crew TCA Viscount 724. An alternative that reduced aisle-width but improved legroom was to fit eight five-abreast seat rows plus two four-abreast. (Vickers-Armstrongs via Author)

way bound for the nearby Wisley test center. Only six days later, it accompanied its Tay Viscount sibling to the Farnborough Air Show, where it gave a stunningly agile flying display, including a memorable low-level pass with three propellers stationary.

The V.700 prototype far more closely patterned the definitive Viscount airliner. It might more accurately be termed a pre-production edition, unlike the proof-of-concept V.630 development prototype. Nevertheless, although maximum takeoff weight had now grown to 50,000 lb., structural adjustments were few. The new model incorporated important changes to the fuel and deicing systems. Outward alterations included a revised cockpit windscreen pattern and a slightly enlarged dorsal fin necessitated by the longer fuselage and also to defeat the V.630's tendency to yaw in choppy conditions. The Viscount's characteristic "petal" engine cowlings made their first appearance, and the main landing-gear members were now offset in the inner nacelles, further broadening a wheelbase already expanded by the wider wing center-section. The maximum cruising speed was raised to 318 mph with the use of upgraded 1,400 shp (1,547 ehp) Dart R.Da.3 Mk 504 engines with increased-capacity compressors. Still-air range reached 970 statute miles with standard reserves and a maximum payload of 53 passengers, or alternatively, 1,200 statute miles with maximum fuel and reduced payload.

The test schedule intensified with two aircraft flying. Although the routines for airframe certification were well established, the novel powerplant took the project into unfamiliar territory. The Rolls-Royce Dart lacked the usual benefit

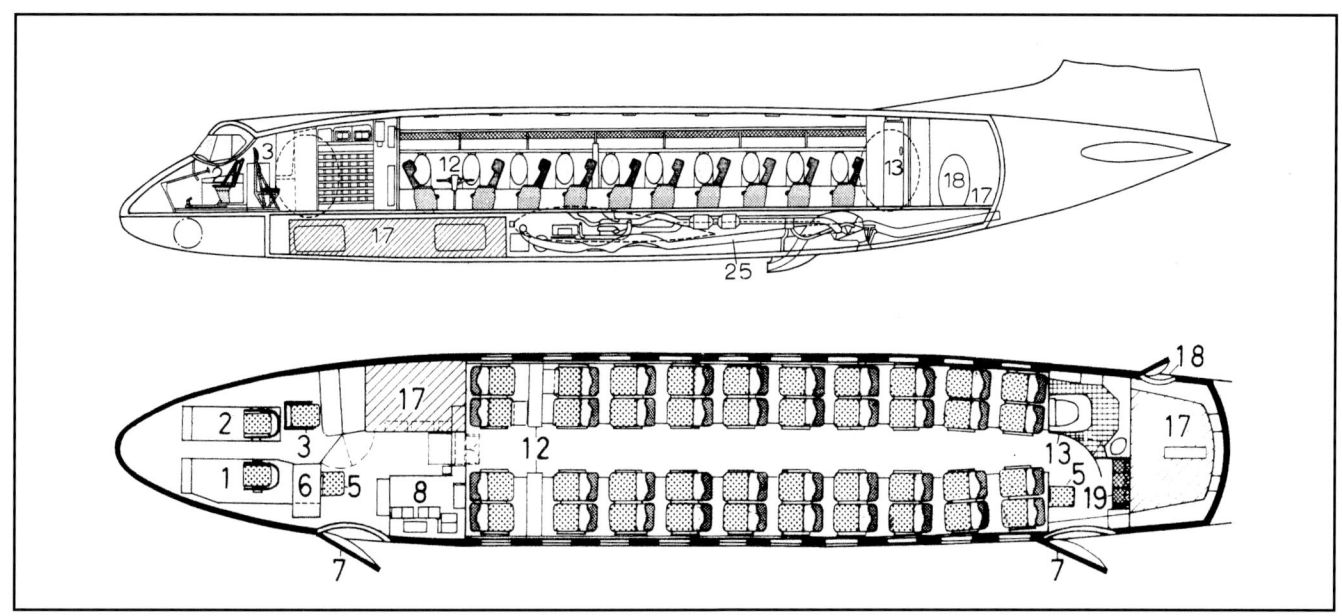

Early Viscount 700 interior arrangement, in plan and lateral views. The layout is for 40 passengers, with each seat row aligned with a window; note the third cockpit seat and the facing pairs of seats at the head of the passenger cabin. (Vickers-Armstrongs via Author)

of proof in military service, leaving much development work to be accomplished. The task was bigger than ratifying the Dart itself. Given urgency by the BEA contract, which anticipated service-entry by early 1953, the capabilities and limitations of the entire turboprop genre needed to be fully explored and defined.

The Dart Dakotas

To facilitate practical experience with the new powerplant, in 1950 Rolls-Royce installed Dart engines in a Dakota (Douglas C-47) on loan from the MoS. However, in order to accelerate engine trials, the MoS also funded the conversion by Field Aviation and Rolls-Royce of two low-time, freighter-configured members of BEA's large Dakota fleet to Dart power. In mid 1951, each received two Dart R.Da.3 Mk 505 powerplants, due to equip BEA's Viscount 701s. Beginning on 15 August 1951, for 18 months the turboprop Dakotas operated both scheduled and *ad hoc* services connecting London (Northolt) with Brussels, Hanover, Amsterdam, Copenhagen, Paris, and Milan. The unpressurized Dakotas cruised regularly at up to 220 mph at 25,000 ft., requiring crews to don oxygen equipment.

A central purpose of these revenue research trials, as BEA termed them, was to determine operating parameters and techniques. The program put the Dart through its paces in a wide variety of conditions, including both hot and cold environments, in order to identify the interplay of all engine-performance factors. Many of these elements, such as jet-pipe temperature, were foreign to the piston-engine world. A concomitant need was to establish accurate means of measuring powerplant output. The relationship of this data to control technique had to be determined and presented usefully to the pilot. During Dart Dakota test flying, the new electric propeller and air-intake deicing system was fitted and verified. The trials also contributed greatly to the refinement of the water-methanol injection process. The Darts themselves performed admirably, although a fatigue failure in the reduction gear in one engine exposed a material weakness and brought delay while Rolls-Royce corrected the defect in all engines. Otherwise, what little trouble there was arose mostly in ancillary systems.

The airline's engineers came to know the Dart engine well. Also, ATC philosophy benefited from experience with the special strictures of turbine-powered airliner practice, especially the need for brief taxi-times and rapid climb and descent. BEA's Dart Dakotas contributed over 3,000 valuable flying hours to Dart research and certification. With their experimental duties complete, they were refitted with their original Pratt & Whitney Twin Wasp piston engines in late 1953. Rolls-Royce's

To establish the characteristics of turboprop operation, as well as to acquire familiarity with the Dart in particular, in 1951–53 two BEA C-47 Dakota freighters – including G-AMDB (c/n 14987) – flew with Dart power on normal scheduled duties. (The A. J. Jackson Collection)

A comparison between the Dart Dakota and its unmodified counterpart reveals how the conversion necessitated a much longer, plumper nacelle. The installation was also larger than would be employed on the Viscount. (The A. J. Jackson Collection)

own Dart Dakota remained at work throughout the 1950s, test flying every variant of the Dart series.

Final Development and Certification

In early 1951, Aer Lingus and Air France declared an interest in buying Viscounts. For most of the year, however, the BEA order for 20 remained the only signed contract. Undeterred, Vickers kept both prototypes hard at work. The V.700 prototype attended the 1951 air shows at Paris and Farnborough. On receiving a restricted C of A on 3 October 1951, it left for tropical trials, starting at Khartoum before transferring to Entebbe and Nairobi for high-altitude work lasting until 30 November. It also undertook demonstrations at Salisbury (Harare), operating services for Central African Airways, and Johannesburg. In the best sense, the entire tropical program was entirely routine, with no problems encountered either in airframe or engine.

On the V.700's return to Britain, Vickers was at last celebrating the Viscount's first export sales. During November, Air France had finally confirmed a 12-aircraft order, and Aer Lingus immediately followed suit, ordering four. The French aircraft, designated V.708, were to seat 49, while the Irish airline specified 53 seats for its V.707s. BEA increased its commitment to 26, but another disconcerting lull would pass before the next order, placed in June 1952 by Trans-Australia Airlines for six 44-seat V.720s. At this stage, a three-crew cockpit was planned for all Viscounts, although this would change for the Aer Lingus aircraft.

A notable feature of development work with G-AMAV in 1951–52 included the earliest trials of a stick-shaker stall-warning device. A vital test took place during further icing trials, based at Prestwick, in March 1952. With airframe deicing switched off, ice was allowed to build up, compromising aerodynamic efficiency until the aircraft reached the beginnings of a stall: the stick-shaker kicked in immediately.

Three months later, now holding a full C of A, G-AMAV embarked on a five-week tour of the Near East, Pakistan and India, lasting until 27 July. In late August, carrying Vickers, BEA and Rolls-Royce personnel, the aircraft departed for Rome, Athens and Nicosia, beginning a program of route-proving over the BEA network lasting into 1953 and adding 550 hours to its logbook. As needed, the airline released the busy V.700 for demonstrations by Vickers, such as that afforded to Aer Lingus at Dublin and Shannon that October.

G-AHRF at Northolt on the morning of 29 July 1950, being readied for departure on the first-ever scheduled departure by a turbine-powered airliner. The British European Airways flight was crewed by BEA personnel. George Edwards is visible on the left of the group of three in the foreground. (ATPH Transport Photos)

The number two propeller starts to turn as G-AHRF prepares to undertake its historic flight. (ATPH Transport Photos)

On 3 September 1950, the prototype V.700 participated in the flying display at the annual SBAC show at Farnborough – its first public appearance. (The A. J. Jackson Collection)

G-AHRF at Northolt during the BEA route-proving exercises of summer 1950. Note the three Vickers Vikings parked in the middle distance – and the passengers apparently walking in front of the Viscount to reach another aircraft. (ATPH Transport Photos)

Amid the heightened effort, the original V.630 continued to contribute. On 27 August 1952, however, while conducting heavy-landing trials at Khartoum for the Air Registration Board, the aircraft's starboard undercarriage failed. Sadly, the resulting airframe injury was too serious to merit repair, and G-AHRF was abandoned. The world's first turbine air transport had completed slightly over 931 historic hours of flying.

Despite this setback, certification flying intensified in anticipation of service-entry. The first production Viscount, a V.701 destined for BEA, took to the air on 20 August 1952. Largely identical to the V.700, it featured Dart R.Da.3 Mk 505 engines (different in detail from the Mk 504 of the prototype, although of similar 1,547 ehp), underfloor holds, further-revised direct-vision windscreen panels, and a cabin accommodating as many as 47. An important, definitive alteration found the powerplants repositioned 18 inches further outboard, in order to mitigate vibration and noise in the forward cabin.

The Vital TCA Order

November 1952 brought arguably the most crucial event in the entire story: Trans-Canada Air Lines announced that it would purchase 15 Viscounts. As a substantial contract for an influential airline with exacting operational requirements, the order – due in no small measure to the efforts of George Edwards, who had vigorously promoted the Viscount in the face of stout rivalry from US manufacturers – was a breakthrough in itself. More significantly – as Edwards himself was never slow to point out – pressure from TCA's operations and engineering teams transformed a Viscount oriented towards the rela-

BOAC Associated Companies – concerned with assisting airlines within BOAC's financial orbit – supported the Viscount strongly once it became a firm commercial prospect. G-AMAV promotionally wore the livery of one such airline, British West Indian Airways, in August-October 1954. Note the updated cockpit windows. (The A. J. Jackson Collection)

tively narrow demands of BEA into an internationally competitive airliner. In particular, not only would the TCA version provide the basis of the bestselling edition of the Viscount 700 series, but the extensive alterations undertaken to optimize the aircraft for Canadian service also created a Viscount of great potential in the US airline market.

Before long, TCA crews had the opportunity to inspect their future workhorse. On 13 February 1953, flown by Jock Bryce with R. J. Baker, chief engineering test pilot of TCA, as second pilot (and with George Edwards as a passenger), the hard-working G-AMAV, now in Vickers house insignia, commenced the first-ever turboprop crossing of the Atlantic, headed for Canada and six weeks of cold-weather evaluation. For appraisal, the airframe had acquired some modifications specified by TCA for low temperature operations; others would result from the trials. The Dart turboprop excelled particularly, demonstrating problem-free starting in Arctic conditions. From TCA's assessment came a requirement for a stronger windscreen and a need to improve the pilot's view, resulting in the alteration of G-AMAV's cockpit window design to a layout destined to become standard for all but the first few Viscount models.

At Last:
The Viscount Enters Service

The V.700 prototype remained at work even after the first Viscount production variants entered service. Its most spectacular mission found it seconded to BEA in October 1953 to participate in the UK-New Zealand Air Race, replacing one of the Corporation's own Viscounts which could not be spared from airline duties. Repainted once more into BEA livery, G-AMAV received the name *Endeavour* and wore the racing number 23 on its fin. With a BEA crew and the airline's chief executive, Peter Masefield, personally undertaking much of the flying, the Viscount left London on 8 October. Its cabin held four supplementary bag-type fuel tanks, bringing AUW to over 64,000 lb. (well beyond its design maximum, and highly taxing on tropical take-offs) and enabling it to complete the 12,500-mile journey to Christchurch

G-AMAV at Heathrow on 8 October 1953, about to depart on the London-Christchurch air race, for which extra fuel tankage raised its weight to almost eight tons beyond design maximum. Although actually the first airliner to reach New Zealand, 'AV attained second place – thanks to a handicap system based on payload – in the race's transport section. Note the blanked-off leading window. In the background stands Britain's other pioneering airliner of the day, the de Havilland Comet 1. (The A. J. Jackson Collection)

in only five sectors. Despite an elapsed time of 40 hrs. 45 mins., on handicap the Viscount came second in its class to a KLM DC-6A. Nevertheless, the demonstration of speed and reliability provided rich publicity, and G-AMAV concluded the adventure with a demonstration tour of New Zealand and Australia followed by a visit to India on the return journey to the UK.

Meanwhile, on 3 January 1953, BEA's V.701 G-ALWE had become the first turboprop aircraft delivered to any airline. At a ceremony on 11 February, Lady Douglas, wife of the airline's chairman, bestowed upon it the name *Discovery*; future BEA Viscounts would each carry the name of a notable explorer. By this time G-ALWE had been joined in the certification and route-proving program by G-ALWF, first flown in mid December. Deliveries of the first several production aircraft to BEA followed rapidly. In early April, G-AMOG (c/n 7) operated some *ad hoc* cargo flights.

The reward for many years of inventiveness, toil and determination was at hand. On 17 April 1953 the Viscount 701 received a full Certificate of Airworthiness, clearing the way for revenue, passenger-carrying operations. One day later, BEA opened the world's first sustained turboprop-powered air service when Viscount G-AMNY (c/n 6), named *Sir Ernest Shackleton*, departed London bound for Rome, Athens and Nicosia. In the cockpit were Captains A. S. Johnson and A. Wilson. The Athens-Nicosia sector operated under charter to Cyprus Airways, who thereby became the world's second Viscount operator. At last, the Viscount was no longer simply a brave and innovative aeronautical project. It had become what it was intended to be: a hardworking breadwinner for the world's airlines.

Full Viscount service began on 18 April 1953. As BEA's fleet grew, Viscount 701s progressively assumed the airline's major routes. An early delivery was G-AMOE (c/n 17), pictured leaving the Heathrow ramp. (ATPH Transport Photos)

VISCOUNT 3 ASSEMBLY

THE VISCOUNT 700 DESCRIBED

Evolving technology and operational experience would not only endow the Viscount with many variations, but would also beget primary sub-types distinguished by extensive revisions. The first Viscount production variant – the V.701 – nevertheless supplied the constructive and dimensional pattern for the entire V.700 series, and provided the fundaments upon which the complete Viscount range would grow.

Fuselage

The Viscount hull was a semi-monocoque, stressed-skin structure fashioned over hoop frames nearly all spaced 20 in. apart. Of Alclad sheet varying in thickness from 18- to 22-gauge, reinforced in high-stress areas such as around openings, the skin was flush-riveted to longitudinal stringers cleated at six-inch intervals around the outside of each frame. Mushroom-headed rivets fastened seams in the pressure cabin. Except for the nose cone, the nosewheel bay, and aft of the rear bulkhead, the entire fuselage was engineered to sustain a pressure differential of 6.5 psi. At the rear bulkhead, a simple, plated structure formed a pressure dome. The nosewheel housing was constructed of stiffened concave sections between braced, pressure-supporting beams.

Each fuselage frame incorporated a crossbeam to bear the cabin floor. Above the baggage hold, forward of the spar, these were reinforced. Aft of the spar, each beam was instead bolstered by a single vertical support. The cockpit floor, of lap-jointed aluminum sheet, extended to the rear nosewheel bay pressure-wall. From there to the rear pressure bulkhead, the flooring – designed to accept a loading of 100 psi, and 150 psi in the baggage areas – consisted of replaceable plywood panels, metal-edged and stiffened with transverse beams. The floor was mostly flush-riveted, but panels along the central aisle were anchored with countersunk screws to enable access to hydraulic and electrical services and the flap motor.

Major sub-assemblies, such as doorframes and the cockpit roof, were fitted before the fuselage segments came together. The unusual control-cabin structure derived from a desire to preserve a circular fuselage cross-section. Since the nose taper began aft of the cockpit area, a conventional pilot's cabin would have meant abandoning this principle. The solution consisted of a separate hood fixed with bolts and rivets to the reinforced edges of a wide, elliptical opening in the hull. Despite methodical fairing into the fuselage contours, the design lent the Viscount cockpit an idiosyncratically domed aspect.

The cockpit windscreens were vinyl laminates, capable of containing the full pressure differential unaided, between semi-tempered glass panes. The famously large, elliptical cabin windows, of which the V.700 series possessed 10 per side, originated in an international airworthiness code setting 26 in. by 19 in. as minimum emergency exit dimensions. Vickers simply adopted these measurements for the window itself, which consisted of a double-pane sandwich separated by a spacer and held in a rubber channel-ring. A desiccator system connected to the gap between the panes removed condensation. In the earliest Viscount 700s every window functioned as an emer-

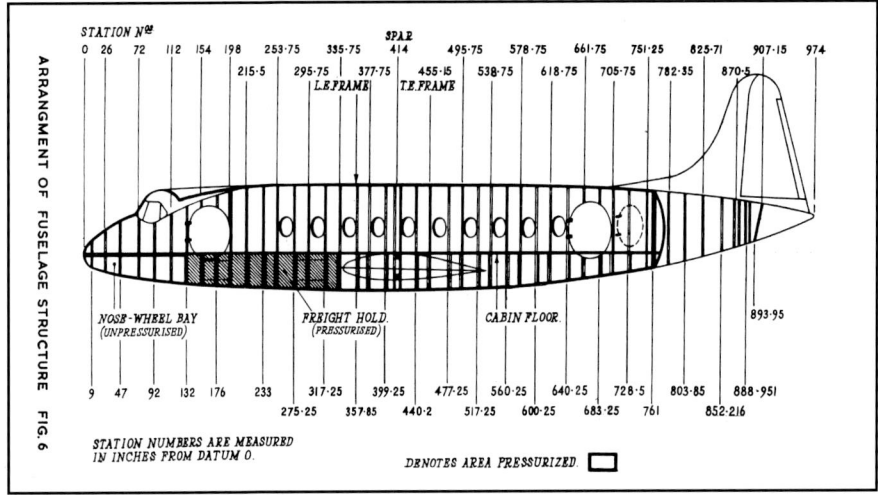

A diagram of the Viscount 700 structure showing station numbers and the positioning of fuselage frames. Note that the area within the cockpit-fuselage fairing remained unpressurized. (Author's Collection)

Early Viscount 700 fuselages completed and ready to join the final assembly line at Weybridge. (Brooklands Museum)

gency exit, three on each side openable from outside.

One cabin door was provided forward of the wing and one aft, both on the port side. In all V.700s, these were outward opening, hinged at the front, and elliptical, measuring 64 in. by 48 in. (forward) and 64 in. by 54 in. (rear). From inside or outside, each main door operated (on release of a safety catch) by means of a handle that actuated 10 peripheral locking bolts. These forced the door outwards against pneumatic seals, inflated from a reservoir in the nosewheel bay to 4 psi above cabin pressure. To starboard, a smaller, elliptical door, also forward hinged and locked by six bolts similarly to the main doors, gave access to the cabin-level rear freight hold. Ahead of the starboard wing, two rectangular doors, each opening out and upwards and measuring 42 in. by 30 in., gave into the underfloor hold.

Wings and Tail Unit

Viscount wing construction followed a Vickers pattern considered unorthodox in that each wing was separate, with no conventional center-section passing through the fuselage, and built around only one spar. Also absent were any typical spanwise stringers. The large, I-section spar lay at 40 percent chord (*i.e.*, at a distance from the leading edge equal to 40 percent of the width of the wing), with two subsidiary, spar-like members at 5 percent and 70 percent chord. Straight and evenly-tapered, the wing possessed a thickness-to-chord ratio of 15.5 at the root and 13.0 at the tip.

Each wing was prefabricated in three sections: an inner wing, bearing the flap assemblies; an outer wing; and a removable tip. After the fuselage sections were united, the inner wings were attached via six tapered bolts fitted to each spar-

At the Hurn factory, this V779D for Fred Olsen is beginning its journey down the production line. The inner wing has been attached to the fuselage; note the visible anti-icing duct in the gap where the nacelle will later be fitted; the nature of the cockpit "dome" is also clear. (Brooklands Museum)

A view of the interior structure of the hull showing the cabin floor support structure. (BBA Collection)

boom root. Four similar bolts joined the auxiliary members to the relevant frames. To bear flight-loads, an integrated beam crossed the fuselage frame between the spars. Flush-riveted fillets faired the wing-skin to the fuselage, although the lower skin ran unbroken from wing to wing. Each inner wing featured nine plate (or diaphragm) ribs, two for each engine mounting and five at flap-rails and similar strategic points. Skin formers of top-hat section, spaced 8 in. apart, served as "mini-ribs." The outer and inner wings met at the outboard engine support-rib: as with the spar-fuselage join, the spar booms fastened together through fork ends and four tapered bolts; four similar assemblies coupled the secondary members.

Save a 16-gauge inboard area, 18-gauge Alclad furnished the wing skin. An uninterrupted sheet ran chordwise from the underside of the forward spanwise member, around the leading edge and rearwards to the upper trailing edge. This detail, aided by the absence of stringers and associated spanwise creasing induced by flight loads, maximized wing efficiency by keeping airflow smooth over an aerodynamically vital area. Flush-riveting and internal butt straps secured the joins in the wing-skin; additional thin panels lined the wing interior. Aft of the trailing edge member, a chordwise extension of riveted stiffened pressings connected the mini-ribs between the flap sections. Removable panels on the undersides offered access to fuel tanks, piping and other items.

Each engine mounting was a braced tubular unit cantilevered from two reinforced wing ribs on the inner wing. The upper parts joined the ribs at the leading edge spanwise member; the lower parts attached both to a bracing strut

Looking forwards inside the bare hull of a Viscount 700: the pattern of circular hoop frames and lateral stringers is clearly shown; straight ahead in the upper portion of the tapering nose section is the oval aperture onto which the cockpit dome will be bolted. (Brooklands Museum)

fixed to the same point and to the spar. At the head of this structure, a firewall incorporated pick-up points for the powerplant.

In structural concept, the Viscount fin and tailplane echoed its wings: each possessed a single, trailing edge spar secured to fuselage frames by two bolts, with a further bolt at the leading edge; a load-bearing beam crossed the airframe between the tailplane spars. Skin-

Taken at Weybridge in spring 1953, this photograph portrays Air France's second and third V.708s. The engines of F-BGNM (c/n 12) (foreground) are in place although neither the outer wings nor the flaps are attached. On F-BGNL (c/n 10), the elevator linkage is just visible where the tailcone will soon be fitted. (Brooklands Museum)

ning was in 20-gauge throughout, flush-riveted. To avoid propeller-wash, the tailplane, with a thickness/chord ratio of 14.0 at the root and 12.0 at the tip, bore a 15-degree dihedral. A dorsal extension to the fin improved longitudinal stability.

Flying Control Surfaces and Systems

All flying-control runs, except those for the flaps, were push-pull rods running underfloor along the port side, passing through roller or spherical guides as appropriate. To enable simpler pressure sealing, devices converted lateral into rotary torque-tube movement wherever runs penetrated the hull. Using materials matching those in the main structure eliminated temperature variation problems. The system embodied autopilot servo circuits: on all V.700s the aileron and elevator drives were beneath the cockpit, while the drive for the rudder was just ahead of the rear pressure bulkhead.

All Viscount control surfaces consisted of metal skin over closely-spaced ribs and D-spar leading edges. The elevators and rudder were each built around a single spar and articulated on four hinges, the bottom rudder-hinge being a vertical pivot seated in the fuselage. To alleviate chafing due to wing flexing, the high aspect ratio ailerons each comprised two sections. An integrated hinge-actuator mechanism operated each aileron. Balanced by normal aerodynamic means, all control surfaces had steel balance weights along their leading edges, with flexible sealing strips to thwart disruptive airflow between fixed and movable surfaces.

Trim tabs for the elevators and rudder functioned via a system of tie-rod and cable circuits. Unlike other control rods, these were affected by temperature and pressure variation and therefore included automatic tensioning devices. The port elevator carried an automatic anti-balance tab outboard and a spring tab inboard, the elevator trim-tab being on the starboard surface. Meanwhile, the aileron tabs were electrically operated on the starboard unit, and of the geared balance type on the port. Rudder, aileron, and elevator locks worked via chains and tie-rods to engage an

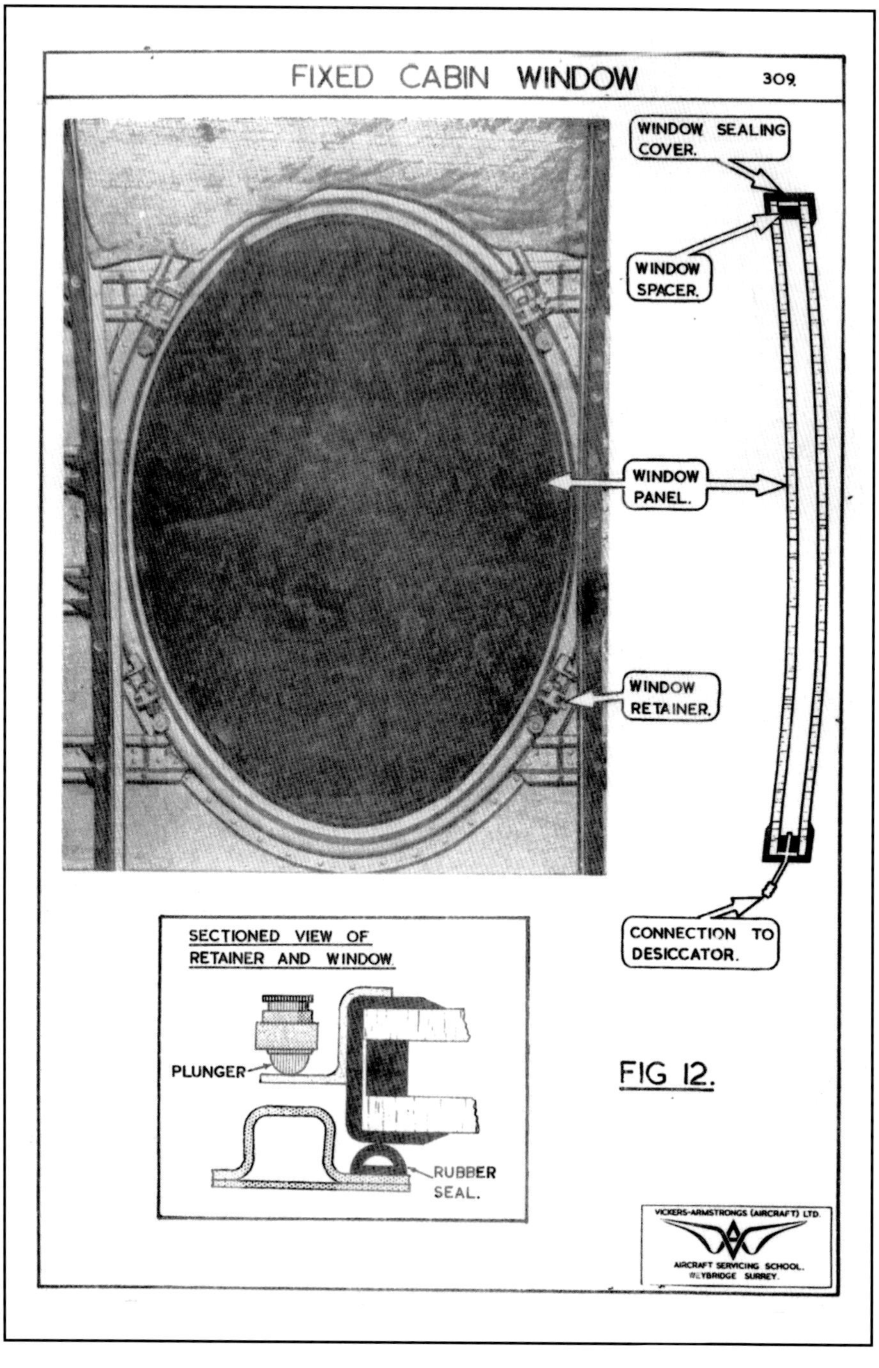

In the earliest Viscounts, all cabin windows functioned as emergency exits; Vickers soon began to fit most windows with fixed units instead. Each window consisted of a double pane, with a desiccator system to eliminate condensation between the panes. (Vickers-Armstrongs via Author)

arm between jaws fixed to the control surface.

Arrayed between wing root and aileron in three sections, the Viscount flap system was a double-slotted design. Each section consisted of main and inner flaps separated by a slot, with another slot between the inner flap and the wing upper trailing edge. As the slots opened, they enhanced effectiveness at lower speeds. An electric motor under the cabin floor drove the flaps via a gearbox, by rotating torque tubes that left the hull through conventional pressure seals to continue spanwise within the trailing edge of each wing. On the tubes, sprockets drove endless chains to slide the flaps in and out on rails mounted on the wing structure. From the wing underside, a vertically pivoting shaft governed the incidence of the flap and acted as a rigid tie-rod at settings of 32 degrees and beyond. Each flap portion was fabricated of a built-up H-section spar, skinned over top hat formers.

Undercarriage

The Vickers-developed undercarriage comprised three forward-retracting, twin-wheeled units. The same strengthened ribs that sustained the inboard engine supports carried the main landing gear. To accommodate the retracted units, these ribs were slightly wider apart. Each main gear leg pivoted about a cross shaft behind the wing spar, braced when extended by a backstay hinged to the trailing edge spanwise member, which also anchored the operating jack. The nose gear and jack were borne on an assembly cantilevered down from the pressure bulkhead at the rear of the wheelwell. This shifted undercarriage stresses away from the cockpit floor. Each unit possessed a single oleo-pneumatic shock absorber strut with

Attaching the inner wing to the fuselage, as seen from the front. (Brooklands Museum)

a 10-in. stroke. That in the main leg contained low- and high-pressure sections, making for greater leg travel and softer landings. A cantilever axle carried twin, cast magnesium-alloy wheels, equipped as required with disc or anti-skid brakes. The nose gear featured castoring and hydraulic steering.

The undercarriage operation involved electrically actuated hydraulic jacks. Extending the jack drove the oleo strut to pivot and push forwards and upwards, forcing the undercarriage up. Shortening the jack tugged the gear down. Both the retracted and extended positions were positive locking. The nose unit incorporated a self-aligning cam to centralize the wheels as the oleo relaxed upon leaving the ground. Mechanically operated doors, hinged

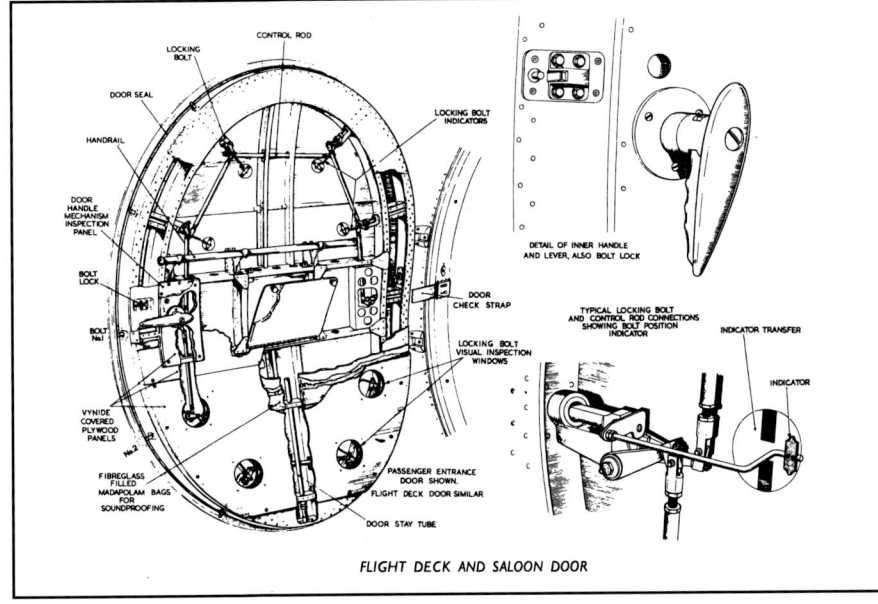

Detail of the oval V.700 cabin entrance door clarifying the locking mechanism; not shown are the pressure seals that were a further part of the system. (BBA Collection)

at the inboard lip of the housing in the rear of the inner nacelles, enclosed the retracted main gear in conjunction with a rear door attached to the backstay. For the nose undercarriage, lateral doors closed via a beam linkage between the front of each door and the oleo strut. With gear extended, all undercarriage doors remained open.

Hydraulic System

The Viscount hydraulic system handled fewer functions than was typical, being responsible for braking, nosewheel steering and undercarriage cycling. Windscreen-wiper operation, although partly hydraulic, involved wholly independent circuits. Several variants featured hydraulically operated airstairs, with hydraulic braking on the port propellers to prevent windmilling during passenger boarding. The hydraulic reservoir, located in the forward luggage bay, gravity-fed two pumps powered by the gearbox of each inboard (originally outboard) engine. In turn, these fed oil to three air-charged accumulators via a common supply pipe and a regulator that kept system pressure between 2,000 and 2,500 psi. A pressure-reducing valve served the steering and brake systems with about 1,500 psi.

To cover for hydraulic pump failure, two of the accumulators, isolated and charged through non-return valves, acted as reservoirs for the duplicated brake system – retaining sufficient pressure for eight brake applications. A manual pump, fed by the emergency hydraulic system from a separate chamber in the main reservoir, provided brake pressure and could lower the undercarriage via an emergency line. Separately, it could be used for ground-testing the hydraulic systems. Nosewheel steering used fluid tapped from the undercarriage down-line to extend horizontal jacks on either side of the nose oleo strut, driving the nosewheel leg round as far as 50 degrees in either direction. A follow-up system returned the valve to the neutral position once the chosen angle had been reached.

A Viscount inner wing, ready for skinning. Note the reinforced spars where the powerplants will be attached. (Brooklands Museum)

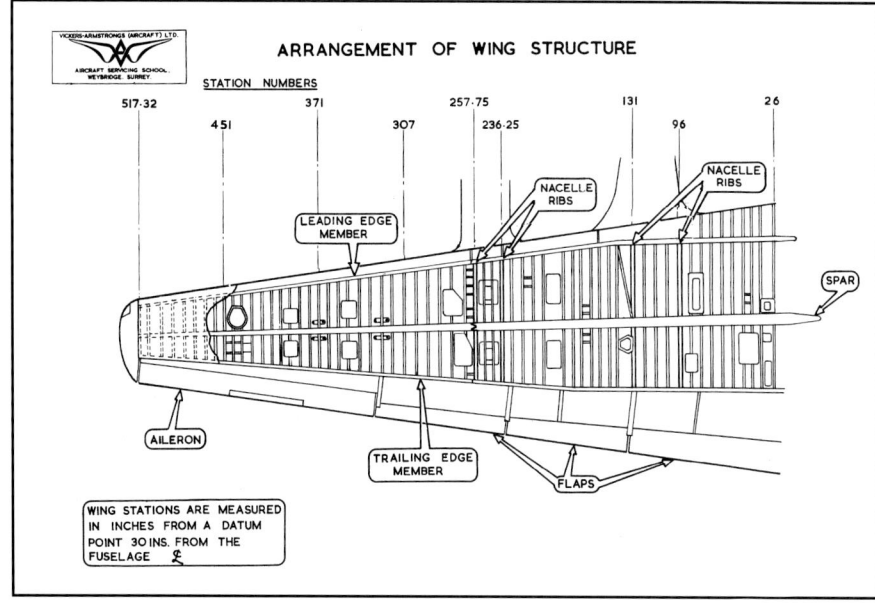

The Viscount wing structure consisted of a single spar, with supplementary leading edge and trailing edge "pseudo spar" members. The diagram also illustrates the pattern of chordwise ribs, reinforced where necessary to sustain the engine nacelles, and the lack of any spanwise stringers. (Vickers-Armstrongs via Author)

Electrical System

Electrical service derived from four 6-kW (later 9-kW) DC generators, one mounted on the accessory gearbox in each engine nacelle, on a negative-earth return system. Carbon-pile regulators controlled output at 27.5 volts, with current supplied through differential relays and interconnected heavy-duty thermal circuit-breakers mounted on the power panel. Fed from the connecting busbars, current passed through a master switch to two 12-volt, 60 amp/hr accumulators connected in series to provide a 24-volt supply. With engines operating, the generators supplied mains fed directly through the circuit-breakers to the control panel in the cockpit, where they branched to the various services.

From the equipment bay in the lower fuselage aft of the spar, four 24-volt 25 amp/hr batteries supplied all normal circuits. Connected in parallel to the battery bus-bar, the batteries held a full charge whenever mains feed remained uninterrupted. Two inverters in the freight hold converted DC to AC power at 115v to operate equipment including the artificial horizon, autopilot and integrated flight system, ADF, turn-and-bank indicators, gyrosyn compass, fuel contents and flow gauges, oil pressure gauges, and engine fire detectors. Electronic actuators operated the low-pressure fuel cock, cross-feed cock, and water-methanol cock. With the other services isolated by inertial crash switches, the batteries also powered the emergency cabin lighting, low-pressure fuel and water-methanol shut-off cocks, and engine and heater fire extinguishers.

Anti-Icing Systems

For removing ice from critical surfaces, such as wing and empennage leading edges, the Viscount eschewed the traditional, repeatedly-inflating rubber "boots" for a hot-air system. From an intake on the inboard side of each inner engine, air passed through a heat-exchanger around which flowed diverted engine exhaust. Thus heated, the air divided into two streams. To supply the outer wing, one stream continued through a duct formed by the gap between the leading edge outer skin and inner plating. A slot in this channel also permitted hot air to flow rearward within the double skin of the wing upper surface, eventually escaping through louvres. The second flow passed through a similar duct inboard and

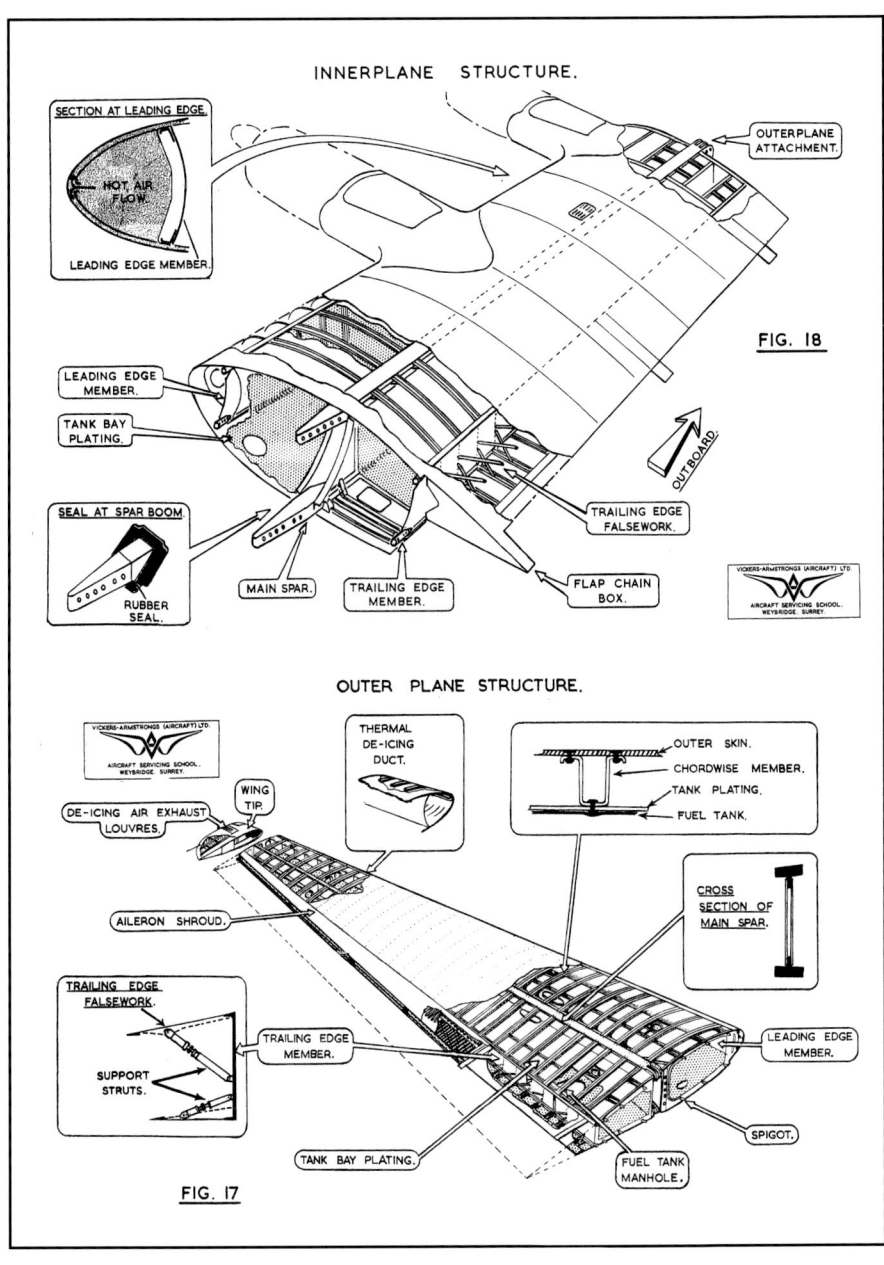

The Viscount wing structure in more detail: note the forked spar attachment by which the inner wing spar slotted into the matching fuselage crossbeam, fixed by six bolts top and bottom; inner and outer wing sections were similarly joined. (Vickers-Armstrongs via Author)

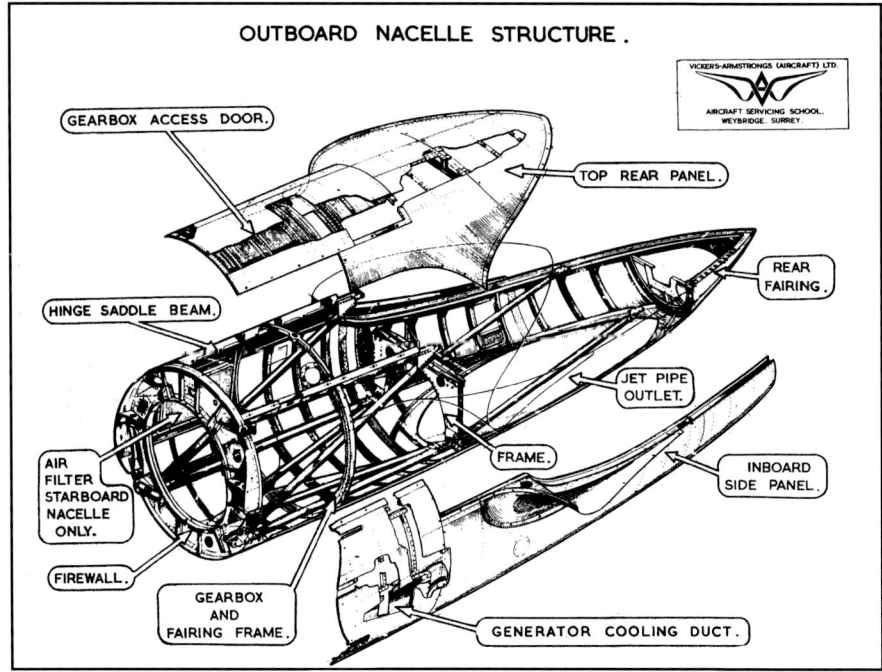

A diagram showing the cantilevered engine mounting structure and aft portion of the nacelle. Not shown are the rearward-opening cowlings, which were articulated from the front of this structure and not attached to the powerplant itself. (Vickers-Armstrongs via Author)

into the pressure hull. The channel then divided, one branch connecting with the system on the opposite side, the other travelling rearward along the fuselage-side below the cabin floor, mirrored by its opposite number. The ducts finally united to pass through the rear pressure bulkhead before branching into three to serve the horizontal and vertical stabilizers. As with the wings, the hot airstream heated these surfaces via passages created by double-skinning. Depending on jet-pipe and outside air temperatures, heat within the thermal deicing channels reached about 150° C in the wings and 90° C in the fin and tailplane – ample for ice preclusion.

Propeller deicing used electric elements embedded in metal-sheathed rubber along the blade leading edges; the spinners were similarly protected. The engine air intake and oil cooler intake were also electrically heated. All of the heating systems operated cyclically, except for the outer lips of the intakes, where heating was continuous. Ingeniously, the engine oil tank formed an integral part of the air intake casing, allowing oil heat to protect the intake-throat. Coil heaters equipped the pitot heads. The system demanded high power, so alternators or AC generators on each engine supplied deicing current independently of the aircraft's main electrical systems. These penalized each engine's output by up to 20 hp.

Windscreen deicing employed ethyl-alcohol, electrically pumped from a 2.5-gal. tank in the cockpit-fuselage fairing behind the crew-cabin, sprayed from the bottom windscreen edge and spread by the wipers. Later variants added electric demisting and deicing using a transparent, electrically resistant material fused to the inner surface of the outer glass pane.

Pressurization and Air Conditioning

Three superchargers mounted on the accessory gearboxes of the port inner and two starboard engines were fed by air arriving through nacelle-mounted filters. To

Joining the powerplant of a Viscount 800 to the inner nacelle structure: of interest is how the engine exhaust section slots into the jetpipe, which is independent of the powerplant. Note also the jetpipe's downward angle. (Brooklands Museum)

achieve the required differential of 6.5 psi, the three systems delivered compressed air at 65 to 100 lb./min into a common conduit. (If necessary, a single supercharger could meet the pressurization needs unaided.) With a rated output of 22 lb./min at 30,000 ft., during pressurization each pulled about 60 hp from its parent engine. There was an underfloor climate-control unit, in which a projecting ventral scoop drew cold exterior air around a heat-exchanger. Air entered the cabin through concealed vents in the lower walls and left through louvres in the roof panels (a path reversed in the V.810). It then passed into the hold areas before finally exiting through two discharge valves on the lower starboard fuselage.

By restricting the rate at which these valves vented, a Normalair regulator maintained interior pressure to that of sea level at up to 15,000 ft. flying altitude, rising to the equivalent of 8,000 ft. at 30,000 ft., replacing cabin air completely every 2.5 minutes. Spill valves in the wings prevented the differential from exceeding 7.5 psi. Loss of a compressor due to engine-failure caused a non-return flap to close in the corresponding duct, stopping cabin air from escaping. To guarantee depressurization on landing, an electric actuator made the outflow valves open fully whenever the nose gear oleo compressed and closed a switch. For unpressurized flight, fresh air entering the belly-mounted intake could be redirected to the cabin through an inlet valve.

Cabin heating traded on the fact that when air is compressed its temperature naturally rises. This trait created sufficient heat to maintain

On the inner powerplant, the jetpipe was angled 40 degrees outward as well as 20 degrees down.

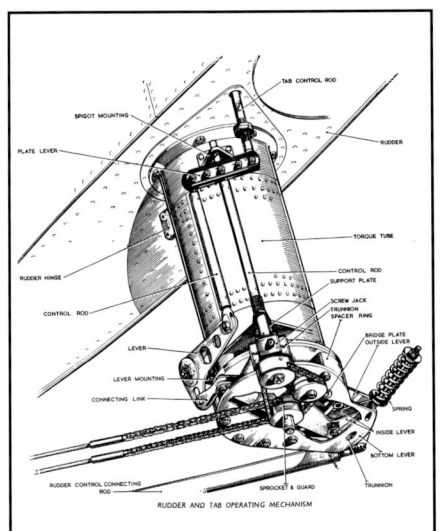

Detail of the rudder and trim tab; as seen here, most flying-control runs were push-pull rods. (BBA Collection)

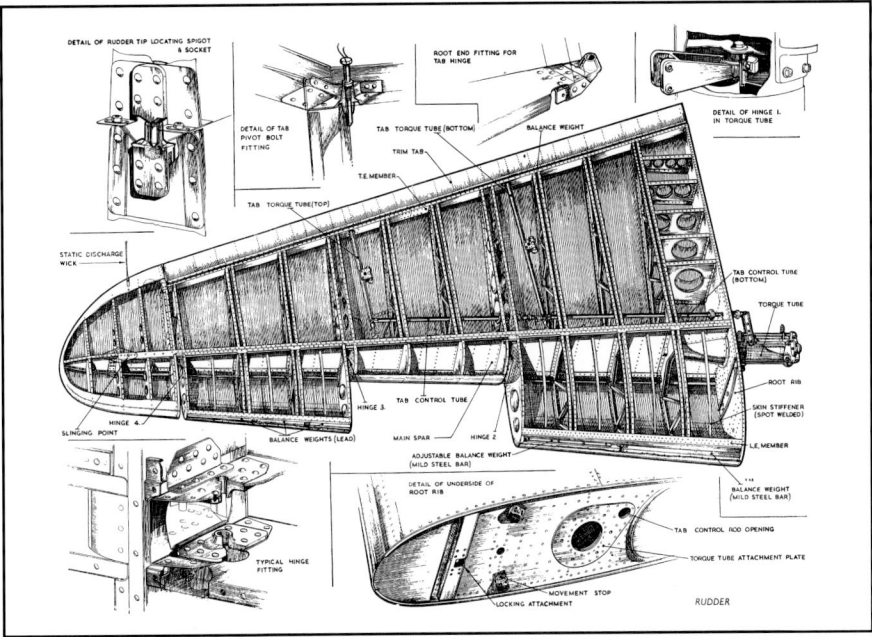

The Viscount rudder structure, with the leading edge towards the bottom of the diagram. Noteworthy are the single spar and the torque tube operating mechanism for the trim tabs. (BBA Collection)

VICKERS VISCOUNT

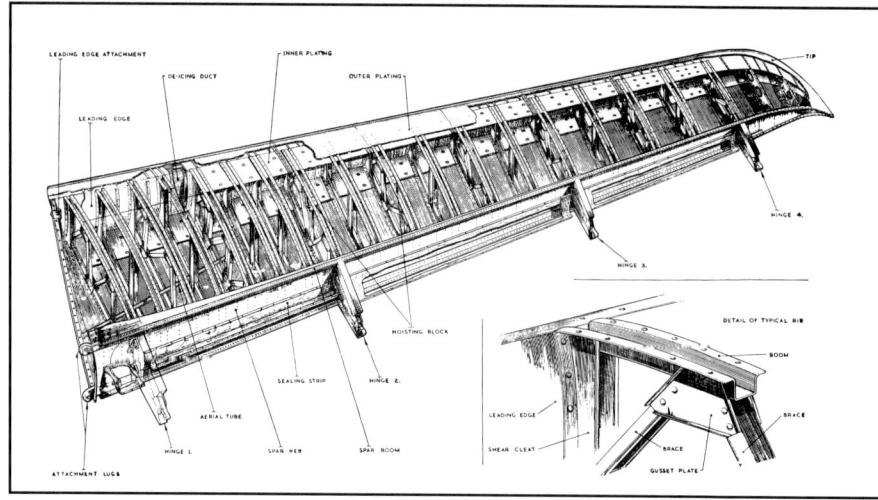

Tailplane construction was built up on a single spar forming the trailing edge; as with the mainplane, a thermal deicing duct ran along the leading edge. (BBA Collection)

65° F for an outside temperature of −50° F, adjustable between 60° and 80° F via the intercooler system and a switch on the flight attendant's panel. Adjustable overhead jets offered passengers further cooling airflow if needed. For service in extreme climates, individual operators specified supplementary refrigeration or combustion heating systems.

Fuel System

Each wing contained two fuel tanks: a main tank, comprising eight interconnected bag cells (four to either side of the spar); and a single-cell secondary tank forward of the spar near the wing root. Each cell, constructed of crashproof, rubberized-nylon fabric, was secured to the wing interior by rubber buttons and individually removable through the panels in the wing undersurface. Use of flexible bags reflected, yet again, the Vickers "play safe" outlook: integral tanks would have cost less weight, but the era's sealing technology was less dependable. To ensure the bag tanks kept their shape, ram air entered through a forward-facing, thermally-heated inlet on the front underside of each outer wing. This provided positive internal pressure, restricted by a relief valve to a maximum differential of 0.5 psi. The same source pressurized the water-methanol tanks.

Early V.700s carried 1,720 Imp. gal. of kerosene – 700 Imp. gal. in each main tank and 160 Imp. gal. in each secondary tank. To raise capacity to 1,950 Imp. gal. in later models, variations in the number and disposition of fuel cells became common, often featuring a further cell in each inboard wing aft of the spar. Some executive-configured Viscounts boasted a 450-Imp. gal. supplementary tank in the freight hold. Viscounts initially possessed no provision for fuel-jettisoning.

Although normally supplied from that on its own side, any engine could be fed from either main tank or both; a cross-feed line connected the port and starboard systems. By way of a cross main, fuel reached the engine-driven pumps from two electric booster pumps located in the inner rear cell of each main tank. Each of these pumps alone was able to ensure maximum RPM from two engines. The secondary tanks did not feed the powerplants directly; instead, via booster pumps, they incrementally replenished the main tanks. Flowmeters in each engine supply line translated the degree of deflection of a vane in the fuel stream into a rate-of-flow reading in lb./hr. Pacitron units in the fuel cells

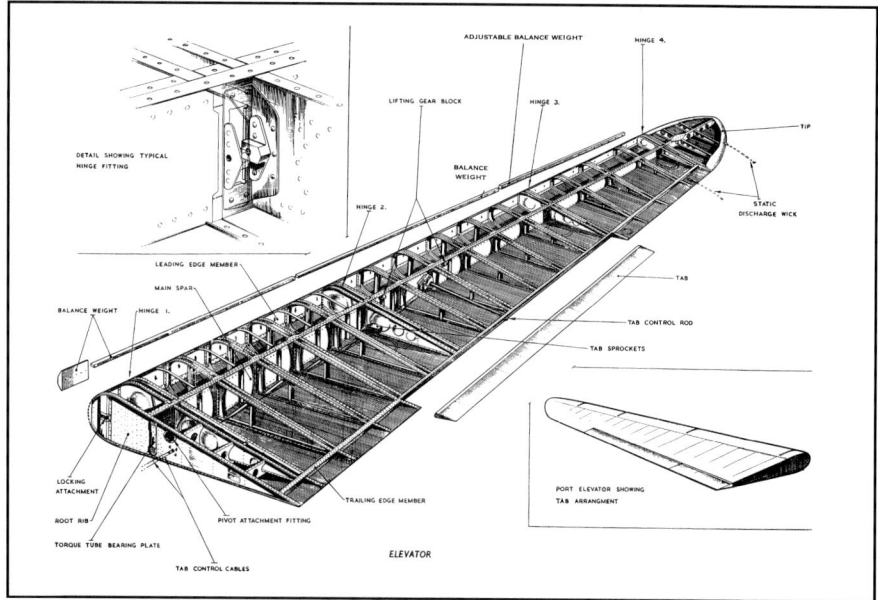

Like the rudder, the elevators were each built around a single spar and attached by four hinges. Note the differing tab arrangements for port and starboard elevators. (BBA Collection)

34

measured fuel quantity electronically, rendering volume into weight. Unaffected by aircraft attitude, the method automatically compensated for temperature-induced fuel-density variations.

Beginning with the fifth V.720, most models included couplings, plumbing and wiring to enable the attachment of two external fuel tanks. Fitted to the underside of the wing leading edge outboard of each outer powerplant, each "slipper" tank could hold 145 Imp. gal. and was simple to attach or remove.

Powerplant and Propeller

The Rolls-Royce Dart was a single-shaft turboprop featuring a two-stage (later, three-stage) axial-flow power turbine, connected directly to a two-stage centrifugal compressor and to the propeller, via a two-stage reduction gear. The oil system, including the tank, was integral. A water-methanol injection installation, fed from a 37.5-Imp. gal. tank in the inner area of each wing, assisted in restoring full power when taking off at high altitude and/or air temperature.

The engine embodied an annular air intake behind the spinner and around the reduction gearing, surmounted by a smaller intake for the oil cooler. Air flowed to the first stage impeller, passing through the rotating guide-vanes, to be diffused by a series of radial passages. Through curved channels formed by the compressor casings, it entered a second, smaller-diameter, impeller. On emerging from the second-stage diffuser, the air had reached around 20 lb./sec and 5.5 times inlet pressure. Seven interconnected "can" combustion chambers, two fitted with igniter plugs, encircled the turbine shaft casting, canted helically to help reduce engine length. Down-

A completed V.810 fuselage awaiting attachment of the cockpit roof shell. (Brooklands Museum)

The hull of a V.748D destined for Central African Airways. Note the more pointed nose-cone, ready for radar installation. In the foreground is a starboard tailplane. (Brooklands Museum)

The fuselage of the first Viscount 803 for KLM – c/n 172, destined to be PH-VIA – travelling outdoors for pressure testing at Weybridge in autumn 1956. (Brooklands Museum)

Major components were removed from the production jigs as early as practicable; aircraft were then held on jacks, as seen in this view in which an inner wing is about to be joined to a fuselage. The radar nose provision (note the hinges) indicates that this is a V.810 variant: the aircraft is a V.838 under construction at Hurn for Ghana Airways. (Brooklands Museum)

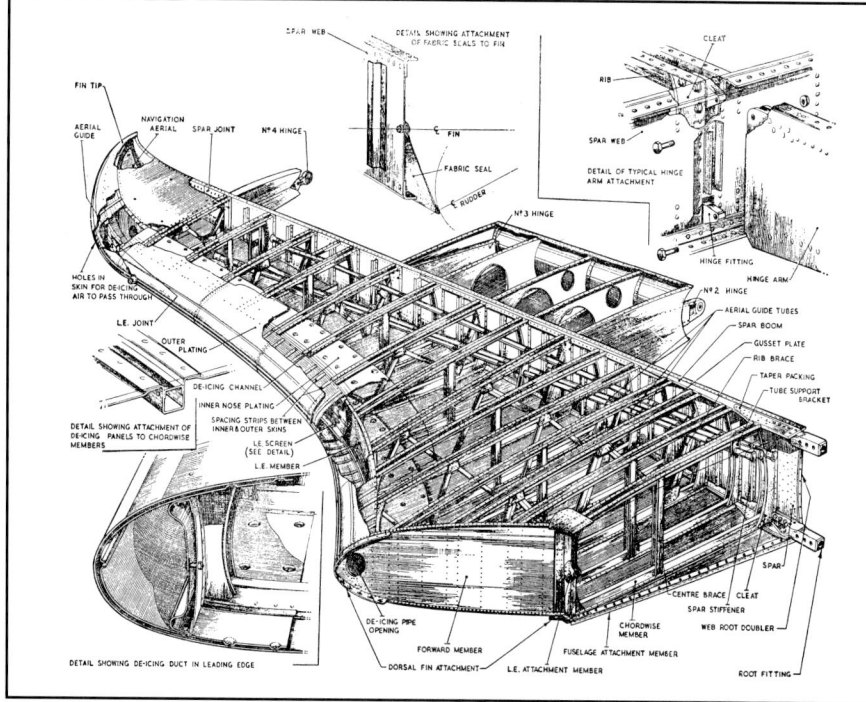

The Viscount vertical stabilizer, illustrating the three-point hinge system (the fourth rudder hinge was seated vertically in the fuselage) and construction detail. (BBA Collection)

stream, the hot gases combined into a single annular flow and coursed through the turbine nozzle blades mounted to the propshaft, thereby spinning the propeller via reduction gearing. Streaming aft, the flow discharged around an exhaust cone into a stainless steel jet pipe – both of which were part of the airframe, not the powerplant – and out to the atmosphere, adding thrust. The jetpipes pointed aft at a shallow downward angle; those inboard were also angled slightly outwards.

Four-bladed Rotol constant-speed propellers, complete with spinners, had been specially designed for the Dart (de Havilland square-tipped airscrews were also available). The small propeller diameter of 10 ft. kept the fuselage and engines near the ground, but resulted in tight ground clearance of 15 in. for the outer propellers and only 7 in. for the inner units. The non-reversing propellers incorporated a variable-pitch mechanism, with a range of 82-85 degrees, consisting of a hydraulic piston linked within the hub to the blade roots.

In early Darts, the propeller gearing achieved a speed reduction to 0.106 percent of engine RPM – a ratio of 0:106. Later variants managed a ratio of 0:093. The Dart became the first British production engine with turbine blades fitted with tip shrouds, a feature enabling a significant reduction in RPM. In an exceptionally long career, the Dart powered many types besides the Viscount, including the Fokker Friendship, Armstrong Whitworth Argosy, NAMC YS-11, and Hawker Siddeley 748. The earliest working Dart was rated at 990 shp; the take-off performance of the version installed in the V.630 prototype was 1,380 ehp. Ultimately, the world's first turboprop powerplant reached a power rating of over 3,400 ehp.

A scene of the final assembly line at Hurn, with a V.720 for TAA nearing completion. A great deal of engine detail can be seen; notice how far the engine is cantilevered forward of the leading edge. (Brooklands Museum)

Destined for CAAC, G-ASDP (c/n 451) receives final attention at Hurn. In China, the aircraft would become B-402, and would eventually conclude its career in Indonesia with Bouraq as PK-IVZ. (BBA Collection)

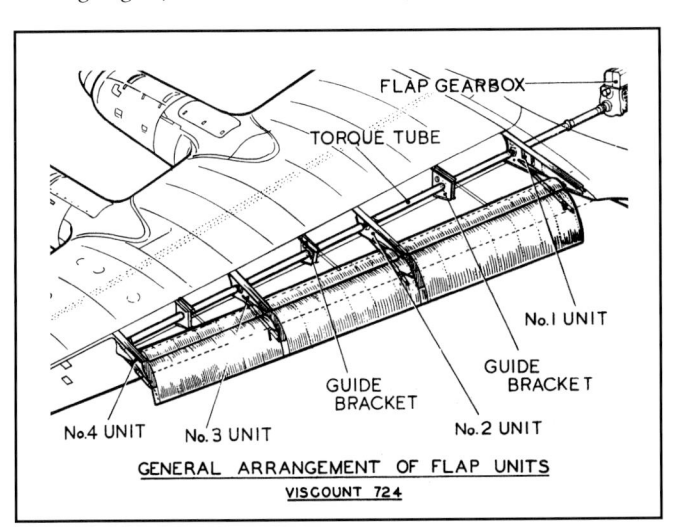

The Viscount's double-slotted flaps were in three sections; a rotating torque tube drove endless chains to slide the flap units back and forth. (Air Canada via Author)

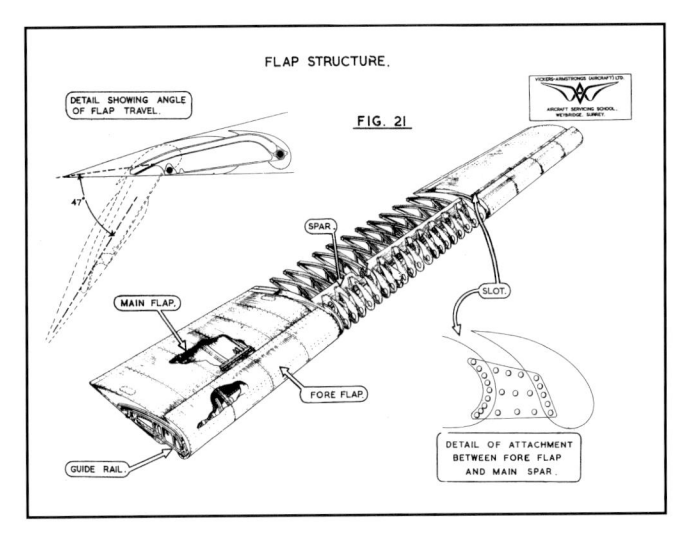

The underlying construction of the double-slotted flaps. (Vickers-Armstrongs via Author)

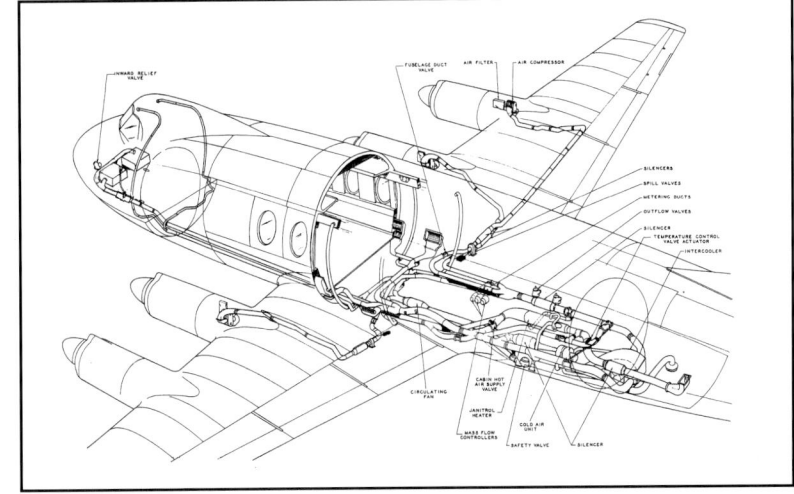

A representation of the Viscount air conditioning system, which used compressors mounted on three of the engines. The V.724 variant illustrated added a Janitrol heater to supply heating while on the ground. (Air Canada via Author)

VICKERS VISCOUNT

FIG. 22 MAIN UNDERCARRIAGE

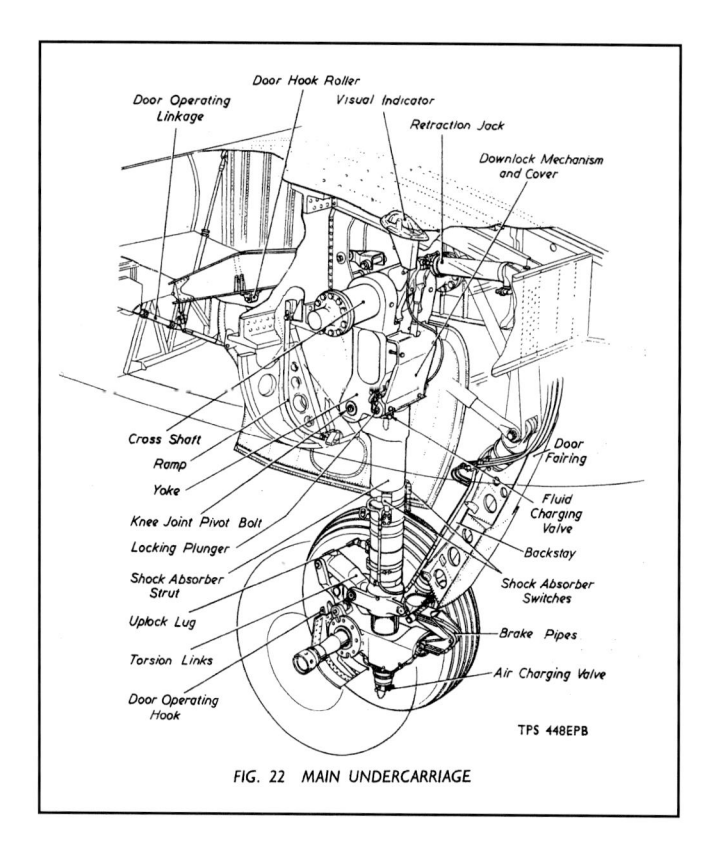

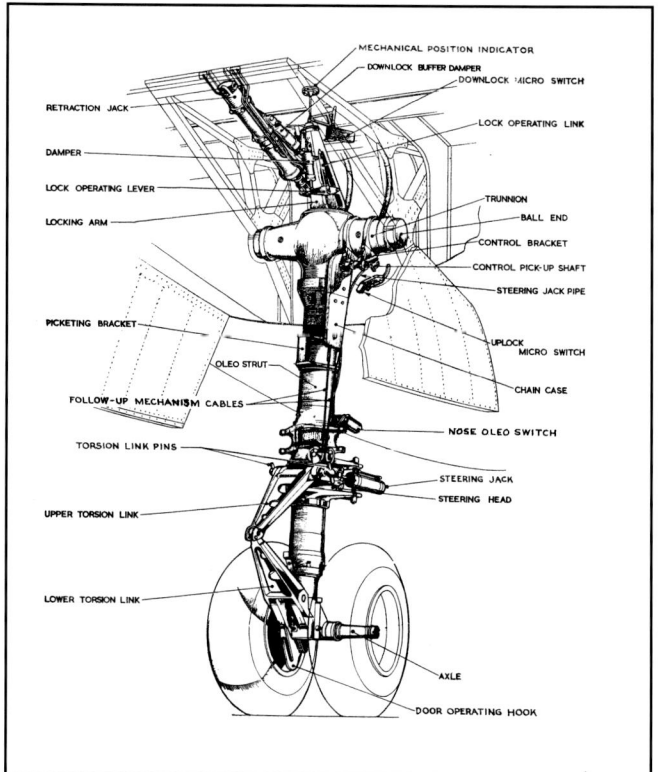

All undercarriage units retracted forwards, which permitted slipstream to assist in emergency extension. Integrated with the main gear shock absorber, microswitches inhibited or allowed the operation of various systems – such as propeller ground fine pitch – according to whether or not the aircraft's weight rested on the undercarriage. (Vickers-Armstrongs)

The nose undercarriage retracted into an unpressurized bay; on leaving the ground, the hydraulically steerable nosewheel centred automatically. To extend the landing gear, the jack at the top of the diagram pushed rearward, pivoting the leg about the trunnion. (Air Canada via Author)

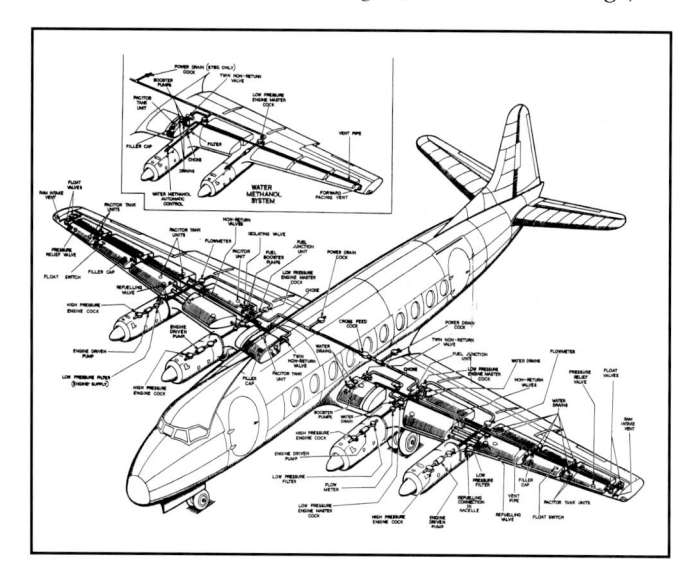

Over time, many different fuel tankage layouts emerged for the evolving Viscount family. Illustrated is the relatively simple system for the early V.701 model. Also shown is the layout of the power-recovery water-methanol system. (Vickers-Armstrongs via Author)

Layout of the hydraulic system, responsible chiefly for braking, nosewheel steering and landing gear operation; flaps were electrically actuated. Note the manual pump for emergency hydraulic operation, located in the forward main-cabin baggage hold. (BBA Collection)

BUILDING ON SUCCESS

DEVELOPING THE VISCOUNT 700

In September 1953, Air France's first Viscounts joined BEA's, with an opening flight over the Paris-Istanbul route. That year, Hunting-Clan, British West Indian Airways, and Iraqi Airways all placed commitments. Moreover, the breed had begun to evolve. The V.707s of Aer Lingus started operations in spring 1954 with a two-crew cockpit – a substantial alteration that prefigured the catalytic Trans-Canada Air Lines version.

Further upgrading appeared that autumn in the V.720 model for Trans-Australia Airlines, which profited from Dart 506 engines of 955 ehp cruising power, enhancing fuel economy and giving a maximum speed of 324 mph. The V.720 introduced a reconfigured fuel system that expanded internal capacity by 230 Imp. gal. and furnished a separate tank, with full crossfeed facilities, to each engine, plus provision for removable external tanks. In the story of Viscount progress, however, the first TCA contract heads the most crucial chapter.

A New Viscount 700: The V.724

Although not an official sub-type, the Viscount V.724 developed for TCA expressed such a great leap ahead of the first V.700 that it merits special attention. Between contract-signing in late 1952 and first delivery two years later, Vickers enriched the basic aircraft with over 250 alterations, many requested by the airline, others necessary for certification. Modified propellers and the 1,547 ehp Dart 506 engine first used in TAA's version (and later retrofitted to most earlier Viscounts), plus the latter's remodelled fuel system, became standard. The heightened power demands led to an upgraded electrical suite. Most importantly, research led to reinforcement of the lower spar boom, increasing wing spar safe life from 17,000 to 30,000 flying hours.

One of the most significant Viscounts built – the first Viscount 724 for Trans-Canada Air Lines, CF-TGI (c/n 40). Canadian authorities required that TCA's Viscounts be finished to US regulatory standards; Vickers also willingly acceded to TCA's many special requests. The result was an aircraft almost ready for the US market – and an invaluable reputation for being customer-oriented. (Author's Collection)

in mind. A team from the US certifying authority, the Civil Aeronautics Administration, visited Vickers early in 1954. That May, the reward arrived. Capital Airlines of Washington, DC, encouraged by free scrutiny of BEA's Viscount results, ordered three Viscounts of the Canadian pattern, designated V.744. The contract included an option – confirmed in August – for no fewer than 37 of a new upgrade, the V.745. At year-end, Capital increased its order by 20, for an astonishing total of 60 Viscounts.

The V.745 debuted a new Rolls-Royce Dart – the R.Da.6 Mk 510 – a major advance over the original R.Da.3 series, the 505 and 506. Out of the work undertaken to produce the TAA, TCA, and Capital variants, Vickers instituted a new standard specification for the V.700, henceforward applying a "D" suffix to the type number of those that were powered by Dart 510s and met US certification requirements. Thus, Capital's new model became the V.745D – and the Viscount V.700D sub-type was born.

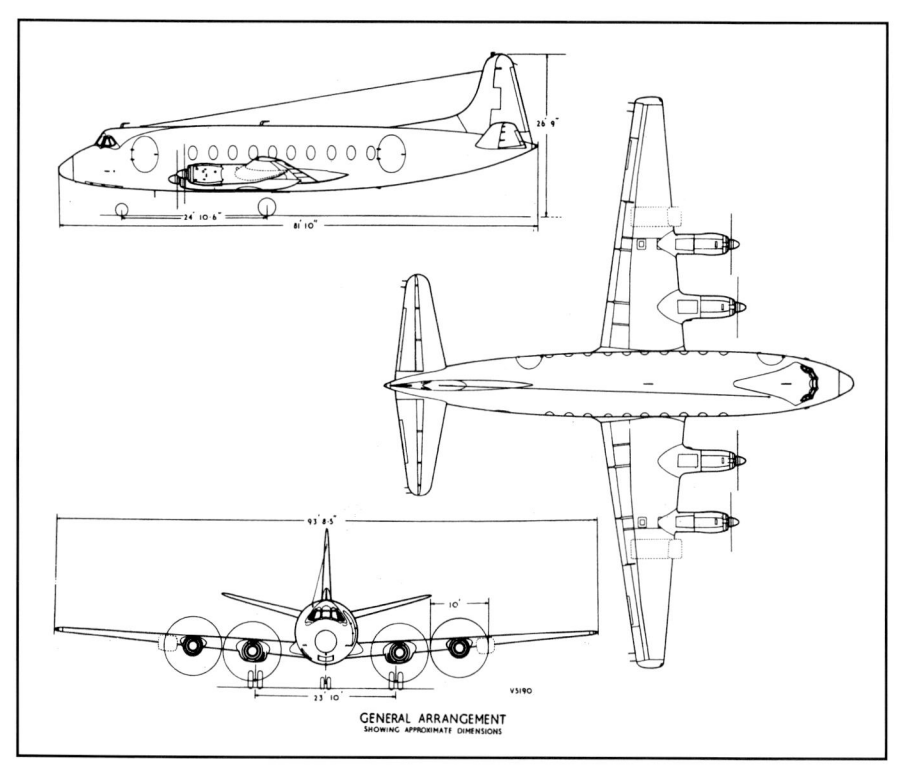

G-AMOO (c/n 28) of BEA was the first Viscount completed on the new assembly line at Hurn. The peripherally-arranged locking bolt visual inspection windows are evident on the inside of each door. (BBA Collection)

This general arrangement view is of the Viscount 748D, and shows not only the revised nacelle profile of the Dart 510s but also the location of the detachable auxiliary wing-mounted "slipper" fuel tanks and the changed nose profile of radar-equipped versions. (Central African Airways via Author)

The basic V.700D – with an AUW of 60,000 lb., a cruising speed of 324 mph, and Dart 510 engines capable of a maximum 1,740 ehp – possessed the best fully-laden range of any Viscount. Gross weight soon rose to 64,590 lb., necessitating 14-in. undercarriage oleos. New paddle-bladed, high-activity trapezoidal propellers boosted performance, assisted by an increased-capacity Dart reduction gear of 0.093 ratio to decrease propeller tip speed. The developed Dart also delivered 1,435 ehp continuous power and 1,025 ehp cruise power. Its larger girth introduced a slightly swollen forward nacelle, replacing the previous straight, cylindrical shape.

For US service, the Viscount – which became the first British designed and built airliner to fly commercially in the USA – underwent a year-long certification process, receiving its US Type certificate on 13 June 1955. Conveniently, 19 of the 26 changes mandated by the CAA already existed in the Canadian variant.

In particular, the TCA-style two-crew cockpit remained. The arduous, short-range nature of Capital's opera-

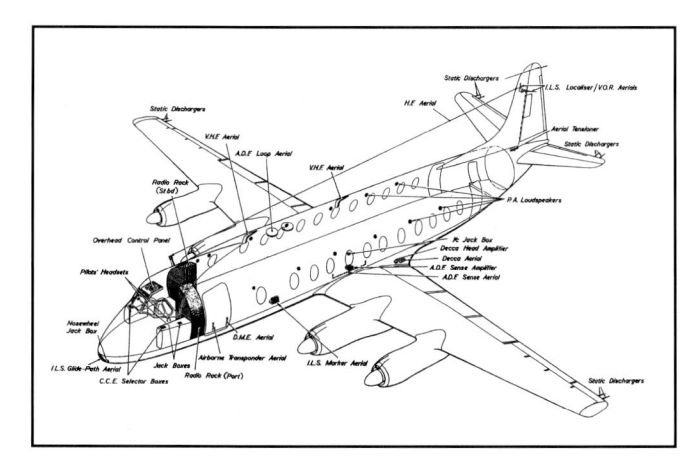

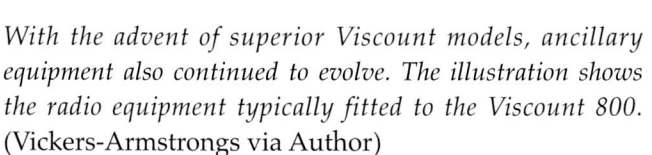

With the advent of superior Viscount models, ancillary equipment also continued to evolve. The illustration shows the radio equipment typically fitted to the Viscount 800. (Vickers-Armstrongs via Author)

Only two models combined the later triangular cockpit apertures with the original production cockpit window layout. One was the V.760D, as evidenced by VR-HFJ (c/n 187) at Kai Tak. (The A. J. Jackson Collection)

tions – more landings per flying hour and more flying at lower, rougher altitudes – necessitated further toughening of the wing spar with redesigned bottom spar booms with 50 percent more area that were constructed of a heavier-duty alloy. Heavier landings accordingly required strengthened undercarriage units. With weather radar now compulsory for all new airliner types under US registry, the nose had to be re-contoured to house the Bendix equipment, adding 8 in. to the fuselage length. Meanwhile, American limitations on maximum landing weights decreed the installation of fuel jettisoning; the system allowed dumping of up to 420 Imp. gal. from each outboard tank via a tube above the outboard edge of the outer flap section. Other requirements focused on instrumentation and performance parameters. Additions also included a Freeon air conditioning system for improved comfort at hot-weather airports, while the V.724's cabin heating was retained. Standard V.700D configuration included provision for auxiliary "slipper" fuel tanks.

Capital also requested about 100 modifications, from an illuminated "return to seat" sign to revised propeller deicing circuits. Customized additions included carry-on luggage racks near the forward entrance and hydraulically operated airstairs (fitted initially to later aircraft in the order) to lessen the aircraft's dependence on ground equipment. For safety during boarding, lowering of the airstairs also activated Dunlop propeller brakes on the port airscrews to prevent windmilling. The airline kept the British pressurization system and Dunlop Maxaret anti-skid braking. However, all instrumentation, navigational and radio equipment, although Vickers-installed, was American.

Although the Capital option pioneered the V.700D, the first firm order came from Central African Airways, in July 1954. With over one third of all production, the V.700D became the most successful Viscount sub-type.

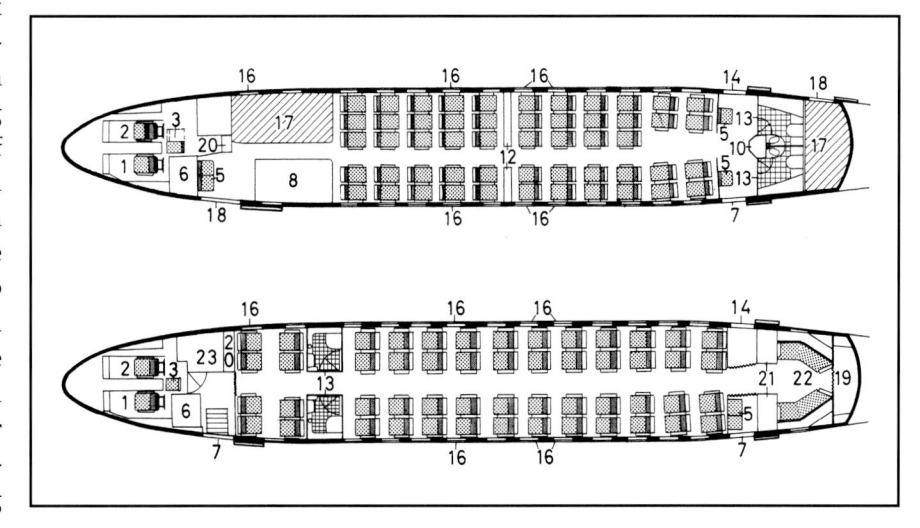

Typical seating plans for the later Viscount models: the economy class arrangement for the V.802 – incorporating expanded freight space at the forward starboard side – shows accommodation for 56; the much more relaxed layout of the V.812 depicts a 44-seater with added lounge accommodation at the rear. (Vickers-Armstrongs via Author)

BUILDING ENOUGH VISCOUNTS

With the Viscount's early success, it grew clear that facilities at Weybridge would be inadequate, particularly as the factory was also needed for the construction of Valiant bombers. Vickers instituted a second Viscount assembly plant, at Hurn, near Bournemouth, erecting two 300-yard production lines and two fuselage assembly halls. In December 1953, beginning with the 25th production aircraft, new Viscounts began leaving the Hurn factory. Vickers relieved further pressure by subcontracting wing fabrication to Saunders-Roe.

Vickers intended to build all future Viscounts at Hurn. However, the order landslide of 1954 rendered even the new 720,000 sq. ft. plant too small. To cure this happy problem, Vickers reopened construction at Weybridge, expanding the factory to cope. The new plan was for Hurn to handle large contracts like the Capital order, leaving miscellaneous orders to Weybridge. When the V.800 came on stream, Vickers re-divided the work – Weybridge concentrating on the new models and Hurn on continuing V.700 production. Later, as V.700 orders tailed off, the two plants shared the task of manufacturing the V.810.

At its height, Viscount construction took place on six separate production lines and assembly shops, jointly measuring over a mile. During 1957, the factories' combined output averaged 10 Viscounts per month – the highest rate ever achieved for a British airliner. Excluding the three V.630/663/700 prototypes and three rebuilds with new c/ns, the construction totals for each factory were as follows:

	Weybridge	Hurn	Total
V.700	39	98	137
V.700D	11	139	150
V.800	67	1	68
V.810	38	45	83
Totals	155	283	438

Stretching the Viscount:
The First Viscount 800 Project

Even before the V.701 entered service, Vickers, prompted by the "tourist class" model then beginning to animate airline strategies, had been working with BEA to formulate a high-capacity Viscount for short-range operations. Grouped under the new generic type number 800, these studies solidified in late 1952 when a new Rolls-Royce Dart, the R.Da.5 series offering 1,690 ehp, surfaced as a prospective powerplant.

Initial V.800 plans propounded a higher-powered V.700 elongated by 13 ft. 3 in. With abridged galley facilities and four more seat-rows, the cabin would have accommodated up to 86 passengers in a coach layout. Oval doors would have been retained. The superior Dart, with gearing altered to give the same propeller speed as earlier engine variants, intimated better fuel economy. Fuel capacity remained at 1,720 Imp. gal.; projected gross weight was 65,000 lb. Interestingly, this was the first private venture Viscount, unassisted by MoS development contracts. At their first V.701's naming ceremony in February 1953, BEA signed for a dozen of the new model, labelled V.801 for the airline, for service in spring 1955, optioning eight more.

Doubt soon clouded the aircraft's performance projections. At only 290 mph, economical cruising speed fell below that of the V.701. Except on sectors of well under 400 miles, payload/range and direct operating costs were threadbare, rendering the important London-Scotland routes scarcely viable and robbing the V.801 of versatility. When deflated traffic forecasts branded the airliner oversized, BEA's commitment dissolved. The first stretched Viscount died unbuilt.

The Definitive Stretch:
The Viscount 800

Within a year, a much more thoroughly thought-out V.800 emerged, embodying a more capable powerplant. The stretched Viscount now possessed a fuselage only 3 ft. 10 in. longer than the V.700's, with the increase entirely ahead of the wing. However, shifting the aft pressure bulkhead rearward by 5 ft. 5 in. allowed internal dimensions extended by 9 ft. 3. in and a passenger cabin 54 ft. long. Power came from the Rolls-Royce Dart R.Da.6 Mk 510 then under development for the V.700D, rated at 1,740 ehp for takeoff. Fuel capacity swelled to 1,940 Imp. gal.

Structurally, the new V.800 reflected the latest sophistications, notably the fortified V.700D wing spars, but added no further changes save a small decrease in tailplane dihedral to 13.7 degrees. The lengthened fuselage allowed an extra window on each side at the rear, plus two on the forward starboard side and one just aft of the forward entrance. Underfloor hold capacity grew from 215 cu ft. to 250 cu ft., offsetting a decrease of 35 cu ft. in rear hold volume due to the rearward cabin expansion. A careful internal redesign widened the cabin by four inches, added improved lighting and other amenities, and

The first non-airline Viscount was a V.737 delivered to the Canadian Department of Transport in March 1955. CF-GXK (c/n 70) was first leased to TCA, complete with titles, to assist crew training. (J. Roger Bentley Collection)

The ability to jettison fuel was mandatory for certification in the United States; 420 Imperial gallons could be shed from each outboard tank via this extensible tube. (Brooklands Museum)

provided better soundproofing. Important cockpit modifications enabled either two- or three-pilot operation. Other up-grades included an electric motor for emergency flap operation.

Arguably, the greatest revision introduced with the V.800 was conceptual. Vickers envisioned the new Viscount as a mixed-traffic airliner, with true passenger and freight flexibility – an early expression of the "Quick Change" genre. Seating able to be folded against the cabin walls became a customer option; additionally, the cabin floor was now all metal, stressed throughout to 150 psi. To facilitate cargo loading, rectangular entrance doors replaced the familiar elliptical units, which had proven troublesome in wind anyway. The remodelled forward entrance offered an exceptional 60 in. of width while the rear door was 27 in. wide; both measured 64 in. high. A starboard service door was added, opposite and identical to the rear passenger entrance. The new doors opened using a parallel linkage system, with the entire unit moving outwards and then shifting laterally to nest against the fuselage.

A maximum takeoff weight (64,500 lb.) identical to that of the V.700D coupled with a higher empty weight inevitably diluted payload and range performance by comparison. Nevertheless, over its intended short-to-medium ranges, the V.800 now exceeded all requirements, enabling high-density seating for up to 71 passengers, or 58 in mixed-class

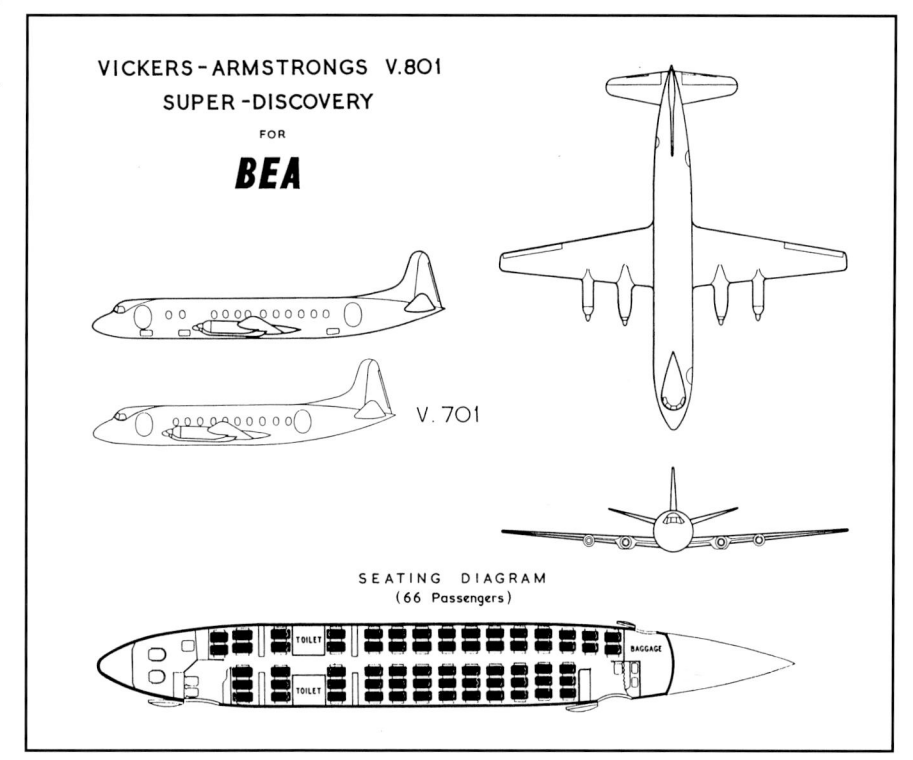

The first stretched Viscount was originally the V.801, planned for 1955. Beside the V.701, the V.801 proposed a 13-ft. 3-in. longer fuselage, seating for 66 (in some layouts, up to 86) and an underfloor hold added aft of the wing, but the same fuel capacity. BEA ordered this high-capacity, short-range Viscount – but the project stalled, supplanted by the economically more realistic V.802 and others of the V.800 family. Other than hull-length, most of the V.801 design, such as door shape, followed that of the V.701. (via R. E. G. Davies)

VICKERS
VISCOUNT

Long-bodied Viscounts featured redesigned entrances, opening outwards to swing sideways against the fuselage. BEA's V.802 G-AOHW (c/n 253) displays the especially capacious forward door of the V.800 sub-type, introduced to permit easy loading of freight. Note the feathered propellers. (Robert Griggs via J. Roger Bentley)

KLM's Viscount interest dated to 1953; Vickers even released illustrations depicting a proposed V.731 in KLM insignia. Instead, the airline waited for the stretched variant, ordering nine V.803s – the first mixed-layout V.800s. PH-VID (c/n 175) excites the attention of air show visitors. (Brooklands Museum)

configuration. Economical cruising speed was 325 mph at 20,000 ft.

In April 1954, BEA converted its V.801 order into a contract for 12 of the redefined variant, with the customer designation V.802 (the airline itself initially referred to it as the Viscount Major), specifying the foldaway seating. With first flight still fourteen months ahead, in May 1955 the airline boosted its purchase to 22. A month later, Vickers booked an important first export order for the longer Viscount when KLM contracted for nine V.803s.

The initial production V.802 (G-AOJA, destined for BEA) served as a prototype, and first flew on 27 July 1956. The latest Viscount received its C of A the following January. Operations began with BEA on 18 February 1957 with four London-Paris return services, soon replacing the Viscount's one-time nemesis, the Airspeed Ambassador, and beginning the phased retirement of the latter from BEA's fleet.

Odd Viscount Out:
The Viscount 806 Hybrid

In January 1956, BEA ordered still more V.800s. Meanwhile, Rolls-Royce had created a radically new Dart, the R.Da.7, distinguished by a three-stage turbine; the engine also featured increased airflow, with two air coolers and twin intakes. Vickers ascertained that the V.800 structure could handle the engine's much greater power, and BEA stipulated the first production variant of the new series, the Mk 520, for its latest Viscounts.

Thus, the intermediate V.806 was born. Supplied only to BEA, it merged the V.800 with the superior engine intended for the recently-announced Viscount 810. The increased shaft horsepower raised takeoff power to 1,890 ehp, boosted climb performance, improved specific fuel consumption, and lifted economic cruising speed. The first V.806, (G-AOYF, in fact designated a

An early purchaser of the V.800 was Transair, an Airwork subsidiary: G-AOXV (c/n 249) was delivered in autumn 1957, fitted with rearward-facing seats for military charter work. On the formation of BUA, this aircraft became possibly the only aircraft painted in BUA's rapidly-abandoned original livery – pale blue cheatline and red fin. (ATPH Transport Photos)

V.806A), flew initially in August 1957. As indicated by the legend *Viscount 806-810* added to its Vickers livery, this aircraft (diverted from BEA's order) was planned also as a certification airframe for the next development, the V.810.

Another Winner: The Viscount 810 Range

The V.800 provided an exceptional short-range airliner. However, airlines wanted a model of similar capacity but that was able to carry a full payload much further. The introduction of the three-stage Dart proved crucial. Around it, Vickers crafted a truly superior airliner, and in 1955, revealed the Viscount 810.

Although unaltered in size, the V.810 was as changed from the V.800 as the latter had been over the developed V.700. At its heart lay the three-stage Dart R.Da.7/1 Mk 525, upgraded from the version installed in the V.806 crossbreed. Capable of 2,100 ehp for takeoff, the Dart 525 was necessarily derated to 1,990 ehp for application in the V.810. The 525 also featured an increased flame temperature and a recalibrated fuel-control unit. Combined with new trapezoidal propellers giving better fuel economy and improved short field performance, the new powerplant enabled a cruising speed of no less than 365 mph.

From an initial AUW of 67,500 lb., the V.810 reached certification at 69,000 lb. Such increases in design weight and speed prompted refinements such as sturdier, longer-stroke main undercarriage units and superior brakes, but also demanded further improvements in airframe fatigue resistance. The ensuing structural revisions became the most extensive in the Viscount's development history, including notably a toughened wing rib and spar structure, with reinforced nacelle and engine mountings to cope with the more powerful engines. Greater power also dictated a strengthened fin and rear fuselage to cope with higher asymmetric loads in engine-out conditions. Local stiffening augmented fuselage strength throughout. To maintain the existing center-of-gravity range and counter diminished effectiveness resulting from greater tail loads, rudder deflection was increased and the elevators received a revised spring tab system. The V.810 would eventually attain a maximum takeoff weight of 72,500 lb. – almost 18 percent heavier than the V.800. Remarkably, structural upgrading contributed only 1,000 lb. of this increase.

Accommodation for weather radar, along with the eight-inch-longer nose shape, became standard, although the equipment itself remained optional. In a retreat from the QC concept, the oversized V.800 forward door shrank in width to 36 in., allowing a second forward cabin window on the port side – the only rapid way of distinguishing a V.810 externally from a V.800. Refining the fit of the rudder and elevator leading edges to the main fin and tailplane surfaces resolved a tendency for snow to collect in the gaps. A major renewal of the electrical system brought a simplified generation circuit, with the engine accessory gearboxes now driving generators of 9 kW DC instead of 6 kW, plus improved batteries and new lightweight wiring throughout. Fuel arrangements included standard provision for 145-gal. slipper tanks, with fuel jettisoning optional.

Among interior alterations, the rear fuselage baggage compartment and its external hatch became optional, with the same area now alternatively serving as a four-seat lounge. The latter variation entailed

Disregarding uninformed Parliamentary criticism that the Viscount was unsuitable for New Zealand operations, the publicly owned New Zealand National Airways Corporation purchased three V.807s – the only V.800s equipped with weather radar during manufacture. ZK-BRD (c/n 281) was the first. (Author's Collection)

Continental Airlines was the launch customer for the top-of-the-range Viscount 810 and first to request the new model's optional four-place rear lounge in place of the baggage compartment. All V.810s possessed a longer, sharper nose to accommodate radar if required. Depicted is N240V (c/n 353). (Author's Collection)

Middle East Airlines' first Viscounts were V.732s leased from Hunting-Clan. The portable fire extinguisher was routine when starting piston engines but would soon be discarded as redundant in the turbine era. (Author's Collection)

a small, elliptical window aft of the rear entrance on each side. Reversing the previous routing, cabin air now entered from adjacent to the light fittings, leaving at floor level. A Freon air-conditioning system remained an option. Also optional was the provision of hydraulically operated airstairs at the forward entrance.

Unfortunately, the V.806A, intended as a supplementary development vehicle for the V.810, was written off in an early landing accident at Johannesburg. Many of the V.810's improvements were nevertheless evaluated in the hard-working V.700 prototype. The first true V.810, G-AOYV, flew initially on 23 December 1957. By this time, the new variant had been gathering orders, starting as early as December 1955 with an important sale to Continental Airlines. The contract was an affirmation of the aircraft's impressive performance projections, as the airline would base its 15 Viscounts at Denver, where aircraft faced takeoff at over 5,000 ft. amsl, high summer temperatures and often an immediate, strenuous climb to clear the Rocky Mountains.

Local Service Viscount: The V.790 Project

In late 1957, Vickers announced the V.790 – dubbed "Local Service Viscount" – founded on the V.700D airframe and conceived as a standardized airliner for short-sector, multi-stop, high-utilization networks typical in the United States.

Capable of five unrefueled, short trips, with full reserves for the final sector, the V.790 promised rapid turnrounds and simple ground handling. Forward airstairs would have been standard, with port propeller braking. A new, low-

Hydraulically operated integral airstairs became an optional feature during production of the V.700D. The mechanism is demonstrated on this Continental Airlines V.812. (Brooklands Museum)

From its beginnings the Rolls-Royce Dart grew in output. The Mk 525 fitted to most V.810s produced 1,990 ehp at takeoff. This view is of the starboard powerplants of a Merpati V.816, PK-RVS (c/n 433), in the climb. (Chris English)

48

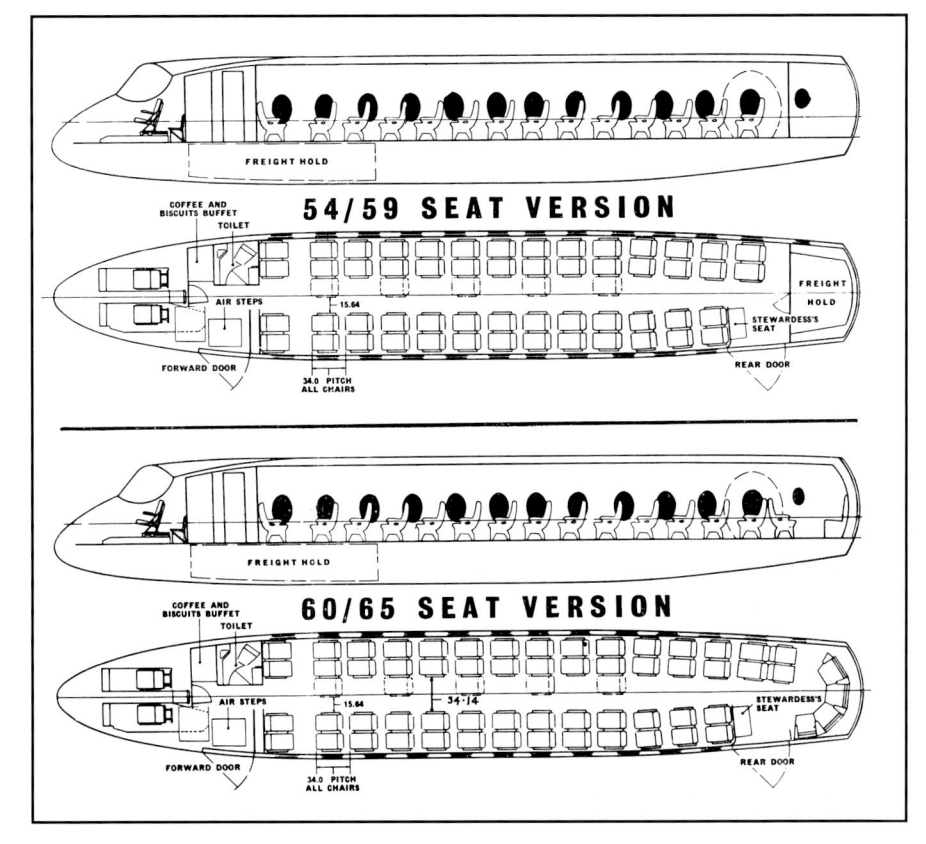

Proposed layouts for the V.790 Local Service Viscount: note that the higher capacity variation included semicircular seating in the cabin rear in place of the hold; both arrangements provided for staggered rows of three seats on the starboard side. Only elementary catering facilities were proposed. (Vickers-Armstrongs via Author)

er-amenity cabin, exploiting the then-new V.800 trim to widen the interior slightly, would have accommodated 65, five abreast, with 54 in a four-abreast layout. For engine restarting at minimally equipped airfields, the starboard outer Dart would normally have kept running, with full starting power guaranteed also by greater battery capacity. Despite utilizing the less powerful Dart 506, the V.790 would have possessed the same takeoff thrust as the V.700D, thanks to the high activity propeller developed for the Dart 510. A higher rate of cycles, heavier landing weights and cruising conditions of 300 mph at 10-12,000 ft. would have meant further structural toughening, particularly in wing reinforcement. The lower typical cruising altitude would have allowed reduction of the pressure differential to 4.5 psi, eliminating two blowers, with lighter pressurization loads compensating for the increased pressurization-cycle rate.

Unfortunately, for its anticipated customers, the V.790 (although cheaper than other Viscounts) remained pricey. In a still-regulated business, small airlines faced trouble raising large sums for aircraft to serve routes they might not be awarded. Serviceable, second-hand piston types abounded. Moreover, the cheaper (and import-duty-free) Fairchild F-27 – a US licence-built, twin-Dart Fokker Friendship – now existed, and, although smaller, elbowed the V.790 aside.

Ultimate but Unwanted: The V.840 Project

Even as Vickers formulated the V.810, Rolls-Royce was breeding a yet more powerful series of Dart turboprop for 1959 availability – the R.Da.11, capable of 2,350 ehp takeoff power and 1,710 ehp in the cruise. From the outset, the manufacturer developed the V.810 airframe to be compatible with this more muscular powerplant, intending to combine them to create a new variant – the V.840. Powered by the proposed

Slipper tanks augmented fuel capacity by 290 Imp gal; most V.700s, plus all V.700Ds and V.810s, possessed the necessary attachment points and plumbing. This is the starboard unit of one of the two V.810s employed by the RAE. (BBA Collection)

VICKERS VISCOUNT

49

Dart R.Da.11 Mk 541 and with operating weights similar to the V.810's, the V.840 promised speeds as high as 400 mph and a takeoff run reduced by nine percent. Range and payload would have been unchanged. Vickers promoted both new models side by side, including a plan to upgrade V.810s to V.840 standard by retrofitting the new Dart once it arrived.

The "400-mph Viscount" almost became a reality. BEA, Iraqi Airways, and the Liberian Government all placed options, and California Eastern evinced strong interest. As the sixties approached, however, the Lockheed Electra and turbojet Caravelle overshadowed the Viscount. Clearly, extracting further sales from Vickers' increasingly well-worn design would take more than higher speed. Quietly, the V.840 fell by the wayside.

Logical Conclusion: Some Final Viscount Projects

Vickers had long eyed the DC-4, DC-6, and L-049 medium haul replacement market. The V.810 and V.840 studies thus begat the V.850, proposing a fuselage extended to 95 ft. and employing the yet more powerful (2,500 ehp) Dart R.Da.8 Mk 550. BEA pondered this "Viscount Major" (reviving its V.801 appellation) as a replacement for its early Viscounts. The V.860 followed, combining the stretched airframe with the Dart 525 of the V.810. Market dynamics, however, pushed the requirements of both manufacturer and airline towards a larger and more adaptable aircraft beyond the ready development potential of the Viscount design, and neither the V.850 nor the V.860 left the drawing board.

A new Rolls-Royce turboprop, the RB.109 (later named Tyne), triggered a two-year-long exploration under the type number 870, initially marrying the new engine to the V.850. The 60-plus V.870 design studies ranged steadily away from the familiar Viscount, contemplating differing wing configurations and even turbojet power. This work led directly toward the much heftier V.900, the first solid design incarnation of the Vickers Vanguard, but separately continued to prompt "Jet Viscount" extrapolations offering various permutations of wing- or rear-mounted powerplants in two- or three-engined layouts. Ultimately, such research re-blended with studies for a jet-powered Vanguard, forming part of the evolution of the VC10 long-range airliner. For the world's first turboprop airliner itself, however, the truth was inescapable. Although four more decades of service awaited, no new Viscount types would see production.

CONSTRUCTING A VISCOUNT

By the time Vickers began producing the V.810, a Viscount took about one year to manufacture, provided specifications were near standard. Special requirements, such as executive interiors, could add weeks or even months to the construction time, mostly in Phase One. Following contract signing, the standard progression was:

Phase One (12 weeks plus): all drawings prepared for the individual variant; manufacturing of parts authorized.

Phase Two (12 weeks): detail manufacturing undertaken for sub and final assembly.

Phase Three (6 weeks): initial sub-assemblies fabricated, including fuselage (built in three sections in separate jigs) and wing portions.

Phase Four (6 weeks): major components assembled, including complete fuselage, inner and outer wings, etc.; completed fuselage pressure tested to 7 psi.

Phase Five (3 weeks): final assembly begun: inner wings mated to fuselage; all flooring installed; electrical conduits added; fin and tailplane attached (sometimes in Phase Six).

Phase Six (3 weeks): outer wings attached; engine nacelles added; rudder and elevators fitted; under-floor systems (including pressurization, anti-icing ducting, air conditioning, main electrics) fitted; landing gear installed; cabin insulation and basic trim attachments incorporated.

Phase Seven (3 weeks): powerplants installed; fuel tanks fitted; ailerons attached; primary cabin elements (galleys, toilets, etc.) installed; cockpit outfitted (instrument panel, central console, radio rack, etc.).

Phase Eight (2 weeks): exterior painted; nacelle fairings fitted; systems checking and calibration begun; flying controls and undercarriage checked for function; electrical system testing begun.

Phase Nine (2 weeks): interior completed (without seating); electrical testing completed; propellers attached; fuel flow checks performed; aircraft removed outdoors for engine running and proof pressurization tests at differential of 8.7 psi; final inspection undertaken; first flight performed.

Phase Ten (2-3 weeks): flight test process carried out and completed; post-flight inspection carried out; UK C of A received; certification received from authority of registration; aircraft formally accepted by customer.

LIFE WITH THE VISCOUNT
THE DAY BY DAY STORY

The Viscount cockpit could scarcely be called large. The dome over the pilots' heads recalled a bomber canopy. Three forward windscreen panels placed a pillar partly in each pilot's line of sight, intensifying the confined impression. In 1953, used to more Spartan amenities, few complained. But, if the flight deck initially impressed with its modern practicality, forty years on, many pilots found its ergonomics challenging.

Either pilot could reach or see most major controls and instruments, but – with engine starting controls, for example, to the right of the co-pilot, and pressurization controls by the pilot's elbow – not always easily. Differing certification and airline stipulations naturally made some Viscount cockpits more convenient than others: both the items mentioned might be on the ceiling panel instead. Straightforward engineering introduced helpful touches, such as a star-wheel the pilot could nudge with his toecap to adjust the rudder pedals. However, accommodating items not originally present could result in, for instance, a radar display all but impossible to see from the right-hand seat. The multiplicity of arrangements precludes a universal description; most of what follows refers primarily to the V.700D and V.800/V.810 models.

The central control panel prominently displayed gauges for jet-pipe temperature and RPM, with torquemeters, outside air temperature gauge and a synchroscope (to assist in RPM synchronization). Below were arrayed engine oil pressure and temperature indicators plus those for cabin pressure. In later versions, oil pressure and temperature were displayed on a single dial, with fuel flowmeters re-positioned beside the other vital indicators. Facing each pilot, a full blind-flying panel carried altimeter, airspeed, turn-and-bank, vertical speed, ILS and other primary indicators. Depending on variant, the coaming panel offered fuel shut-off and feathering controls and fire-fighting switches. On North American and some later models, these were on the ceiling panel, together with items found, in other variants, mostly on side panels – such as deicing, lighting, pressurization, heating and radio controls.

For engine starting, external power could be plugged into a socket – depending on variant, in either the aircraft belly or the outer starboard nacelle – that operated a relay circuit to offload the generators and isolate the batteries. In most models, starting normally required a 1,500-amp supply; if necessary, the aircraft's batteries could be paralleled with a lower-strength external unit to achieve the required power. To avoid undue load, a cut-out prevented starting unless the propeller blades were in zero pitch. Before starting, fuel trim was set and fuel booster pumps switched on and checked for adequate pressure; the master switch would then be activated and the engine start selector positioned to the relevant powerplant. In the engine, igniters lit the third and seventh combustion cans, interconnectors spreading combustion to the other chambers. Managing starting could

Like the passenger entrances, rear baggage hold doors on all V.700s were oval and opened forwards. Note that G-ANHB (c/n 62) has an eleventh cabin window, indicating that it has received BEA's high-capacity conversion, enabling it to seat 63 passengers. (ATPH Transport Photos)

> ### Viscount Cockpit Window Variations
>
> Vickers progressively modified the Viscount cockpit window pattern. This unofficial list notes the distinguishing features and the Viscount variants associated with each pattern.
>
> **Type 1:** Side (direct vision) window divided laterally into 3 even panels. (V.630, V.663).
>
> **Type 2:** Larger direct vision window with single pane only (prototype V.700 initially).
>
> **Type 3:** Interim version of definitive pattern (type 6), lacking small rear paneless (prototype V.700 as modified; V.732, V.739 [not V.739A or V.739B]; V.720 as modified; 1st production V.701 for trials in 1952, refitted with type 4 pre-delivery).
>
> **Type 4:** As type 2, but direct view panel split laterally into 3 uneven panels (V.701, V.702, V.707, V.708, V.720 as delivered, V.723, V.730, V.735, V.747, V.773).
>
> **Type 5:** As type 4 but with additional triangular panels (V.734, V.760D).
>
> **Type 6:** Definitive pattern; as type 3 but with additional triangular panels (all other variants).

be tricky – especially with the first engine, unassisted by generators on already-running engines – since a slow start could result in too rich a fuel/air ratio and a JPT above the permitted maximum of 640° C.

Crews converting from piston-engined types found the time lag between engine control inputs and their effect novel. For each powerplant, a single lever (analogous to a piston aircraft's throttle, and often referred to as such) on the pedestal between the pilots governed both propeller pitch and fuel feed – although the latter only at engine speeds over 10,000 rpm. Crews therefore had a delicate throttle control task at lower settings (in effect, while maneuvering on the ground) in order to avoid unbalancing the mixture ratio; unsurprisingly the recommended taxiing method was to avoid differential power and instead use nosewheel steering and a constant power setting. The most critical powerplant factor was the turbine inlet temperature, as indicated by JPT: this had to be maintained within limits by using the fuel trimming control, which varied the relationship between fuel flow and engine speed. Ahead of takeoff, each engine would be calibrated by trimming the JPTs at 12,000 rpm to an air-temperature-related fuel datum value which – given a constant OAT – would remain valid for all altitudes and power settings. En route fuel-flow re-trimming frequently proved necessary; trim was always rechecked before landing.

To prevent wind from loading the control system of a parked aircraft by deflecting rudder, ailerons or elevators, all of them could simultaneously be locked in the neutral position, obviating the need for external, detachable gust-locks. These integral locks could resist tail wind gusts of 100 mph. Taxiing was undertaken with the locks engaged. For safety, a mechanism prevented the power levers from being opened further than one third (*e.g.*, for takeoff) unless the control surface locks had been released by means of a handle on the pedestal. Similar safety devices included an airspeed pressure switch to preclude undercarriage retraction at speeds below 80-85 kt.

Anti-icing – due to the power demand, not permitted on the ground – could be switched on immediately before takeoff if required. Normal takeoff flap setting

BEA eventually altered nine V.806s to V.802 performance standards by replacing their Dart 520 engines with Mk 510s. Viscount 806 G-AOYM (c/n 262) is at Heathrow in the 1960s; the Vauxhall Cresta lends a period touch. (ATPH Transport Photos)

KEY TO COCKPIT LAYOUT

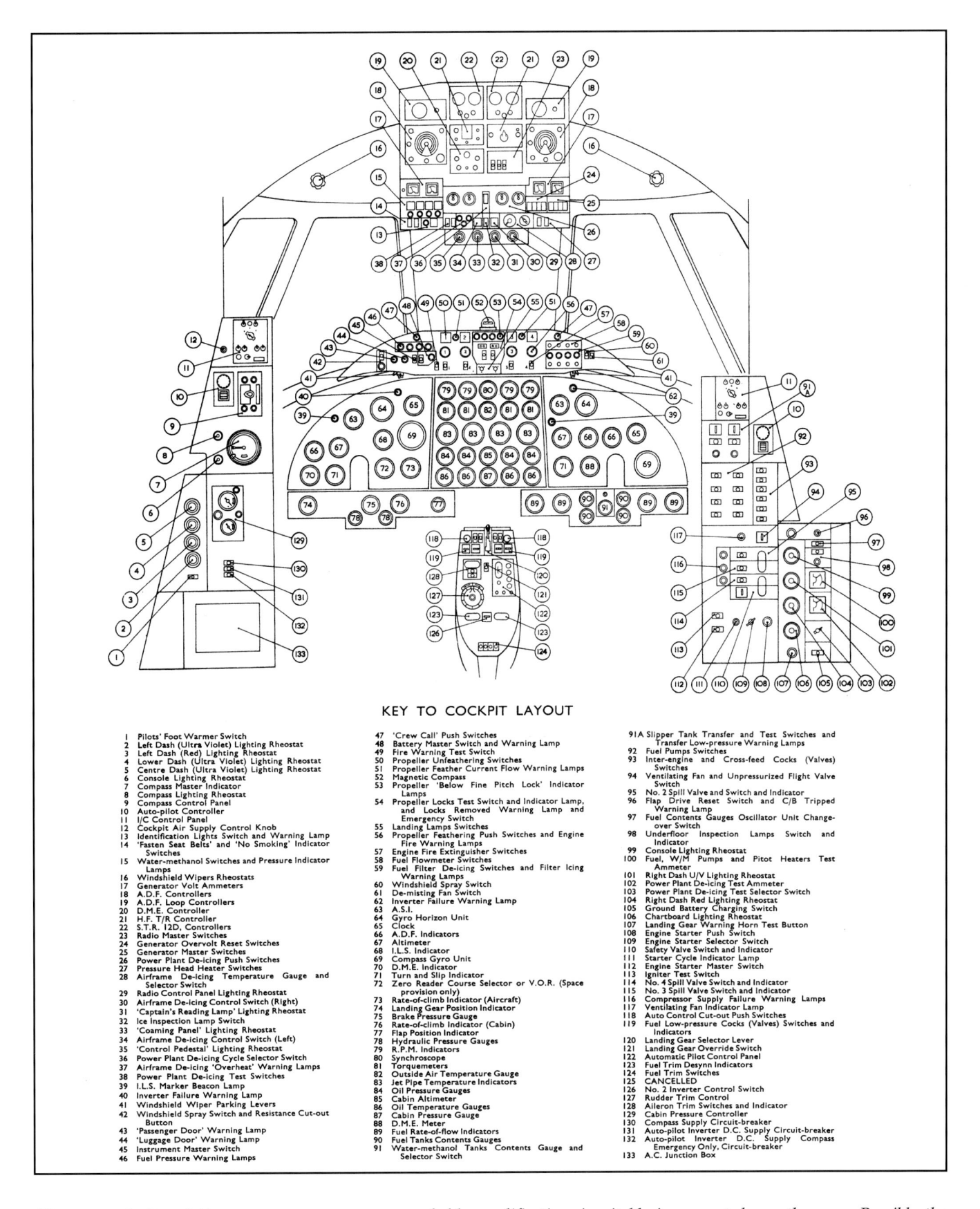

1. Pilots' Foot Warmer Switch
2. Left Dash (Ultra Violet) Lighting Rheostat
3. Left Dash (Red) Lighting Rheostat
4. Lower Dash (Ultra Violet) Lighting Rheostat
5. Centre Dash (Ultra Violet) Lighting Rheostat
6. Console Lighting Rheostat
7. Compass Master Indicator
8. Compass Lighting Rheostat
9. Compass Control Panel
10. Auto-pilot Controller
11. I/C Control Panel
12. Cockpit Air Supply Control Knob
13. Identification Lights Switch and Warning Lamp
14. 'Fasten Seat Belts' and 'No Smoking' Indicator Switches
15. Water-methanol Switches and Pressure Indicator Lamps
16. Windshield Wipers Rheostats
17. Generator Volt Ammeters
18. A.D.F. Controllers
19. A.D.F. Loop Controllers
20. D.M.E. Controller
21. H.F. T/R Controller
22. S.T.R. 12D, Controllers
23. Radio Master Switches
24. Generator Overvolt Reset Switches
25. Generator Master Switches
26. Power Plant De-icing Push Switches
27. Pressure Head Heater Switches
28. Airframe De-icing Temperature Gauge and Selector Switch
29. Radio Control Panel Lighting Rheostat
30. Airframe De-icing Control Switch (Right)
31. 'Captain's Reading Lamp' Lighting Rheostat
32. Ice Inspection Lamp Switch
33. 'Coaming Panel' Lighting Rheostat
34. Airframe De-icing Control Switch (Left)
35. 'Control Pedestal' Lighting Rheostat
36. Power Plant De-icing Cycle Selector Switch
37. Airframe De-icing 'Overheat' Warning Lamps
38. Power Plant De-icing Test Switches
39. I.L.S. Marker Beacon Lamp
40. Inverter Failure Warning Lamp
41. Windshield Wiper Parking Levers
42. Windshield Spray Switch and Resistance Cut-out Button
43. 'Passenger Door' Warning Lamp
44. 'Luggage Door' Warning Lamp
45. Instrument Master Switch
46. Fuel Pressure Warning Lamps
47. 'Crew Call' Push Switches
48. Battery Master Switch and Warning Lamp
49. Fire Warning Test Switch
50. Propeller Unfeathering Switches
51. Propeller Feather Current Flow Warning Lamps
52. Magnetic Compass
53. Propeller 'Below Fine Pitch Lock' Indicator Lamps
54. Propeller Locks Test Switch and Indicator Lamp, and Locks Removed Warning Lamp and Emergency Switch
55. Landing Lamps Switches
56. Propeller Feathering Push Switches and Engine Fire Warning Lamps
57. Engine Fire Extinguisher Switches
58. Fuel Flowmeter Switches
59. Fuel Filter De-icing Switches and Filter Icing Warning Lamps
60. Windshield Spray Switch
61. De-misting Fan Switch
62. Inverter Failure Warning Lamp
63. A.S.I.
64. Gyro Horizon Unit
65. Clock
66. A.D.F. Indicators
67. Altimeter
68. I.L.S. Indicator
69. Compass Gyro Unit
70. D.M.E. Indicator
71. Turn and Slip Indicator
72. Zero Reader Course Selector or V.O.R. (Space provision only)
73. Rate-of-climb Indicator (Aircraft)
74. Landing Gear Position Indicator
75. Brake Pressure Gauge
76. Rate-of-climb Indicator (Cabin)
77. Flap Position Indicator
78. Hydraulic Pressure Gauges
79. R.P.M. Indicators
80. Synchroscope
81. Torquemeters
82. Outside Air Temperature Gauge
83. Jet Pipe Temperature Indicators
84. Oil Pressure Gauges
85. Cabin Altimeter
86. Oil Temperature Gauges
87. Cabin Pressure Gauge
88. D.M.E. Meter
89. Fuel Rate-of-flow Indicators
90. Fuel Tanks Contents Gauges
91. Water-methanol Tanks Contents Gauge and Selector Switch
91A. Slipper Tank Transfer and Test Switches and Transfer Low-pressure Warning Lamps
92. Fuel Pumps Switches
93. Inter-engine and Cross-feed Cocks (Valves) Switches
94. Ventilating Fan and Unpressurized Flight Valve Switch
95. No. 2 Spill Valve and Switch and Indicator
96. Flap Drive Reset Switch and C/B Tripped Warning Lamp
97. Fuel Contents Gauges Oscillator Unit Change-over Switch
98. Underfloor Inspection Lamps Switch and Indicator
99. Console Lighting Rheostat
100. Fuel, W/M Pumps and Pitot Heaters Test Ammeter
101. Right Dash U/V Lighting Rheostat
102. Power Plant De-icing Test Ammeter
103. Power Plant De-icing Test Selector Switch
104. Right Dash Red Lighting Rheostat
105. Ground Battery Charging Switch
106. Chartboard Lighting Rheostat
107. Landing Gear Warning Horn Test Button
108. Engine Starter Push Switch
109. Engine Starter Selector Switch
110. Safety Valve Switch and Indicator
111. Starter Cycle Indicator Lamp
112. Engine Starter Master Switch
113. Igniter Test Switch
114. No. 4 Spill Valve Switch and Indicator
115. No. 3 Spill Valve Switch and Indicator
116. Compressor Supply Failure Warning Lamps
117. Ventilating Fan Indicator Lamp
118. Auto Control Cut-out Push Switches
119. Fuel Low-pressure Cocks (Valves) Switches and Indicators
120. Landing Gear Selector Lever
121. Landing Gear Override Switch
122. Automatic Pilot Control Panel
123. Fuel Trim Desynn Indicators
124. Fuel Trim Switches
125. CANCELLED
126. Inverter Control Switch
127. Rudder Trim Control
128. Aileron Trim Switches and Indicator
129. Cabin Pressure Controller
130. Compass Supply Circuit-breaker
131. Auto-pilot Inverter D.C. Supply Circuit-breaker
132. Auto-pilot Inverter D.C. Supply Compass Emergency Only, Circuit-breaker
133. A.C. Junction Box

Viscount cockpit variations were numerous, compounded by modifications inevitably incorporated over the years. Possibly the most representative was that of the V.700D: the diagram is for a V.748D in operation in the early 1960s. (Central African Airways via Author)

VICKERS VISCOUNT

The overhead panel in this V.800 cockpit carries engine starting and fuel control switches, plus radio and airframe anti-icing controls. Variations in cockpit configuration were numerous. (Brooklands Museum)

was 20 degrees (43 percent). Following brake-release and application of full power, ground steering continued as speed built to 70-80 kt, when the rudder took effect. After decision speed (V_1), in normal conditions rotation (V_R) would be initiated at about 3-4 kt below takeoff safety speed (V_2). Takeoff technique in a crosswind or turbulence emphasized rudder-assisted steering and keeping the nosewheel firmly in contact until V_2. The maximum allowable crosswind component was 27 kt.

For takeoff at high elevations and in high ambient temperatures – where lower air density brings less efficient combustion and higher JPT by increasing the fuel/air ratio – the engines incorporated a water-methanol injection system. To cool the air and increase its density, the system sprayed a mixture, automatically metered according to ambient air pressure, of pure water and methanol (in the ratio 63:37 by weight at 60° F) into the airflow entering the first-stage compressor. This restored full performance for takeoff and initial climb, also granting a margin for dealing with an overshoot or engine failure. The electrically pumped supply provided enough for three one-minute bursts of simultaneous maximum power in all engines. Water-methanol injection was routine in Dart 505, 506, and 510 engines in temperatures above ISA (15° C at sea level). The Dart 520, 525 and 530 used in the V.806 and all V.810 models, however, were derated for takeoff, and required water-methanol boost only at above ISA plus 6° C (21° C).

Initial climb would be made at V_2 until reaching the calculated flaps-up climb speed; on retraction of the flaps, power would be reduced to climbing RPM. As appropriate, crew duties during the climb included beginning fuel transfer from secondary to main tanks to shift weight outward and dampen wing-flexing.

Early Viscount crews often described its flight controls as responsive, with low stick forces. Later pilots considered them heavy, especially in the V.800/V.810. In fact, control runs required 25-35 lb. of muscle-power, depending on conditions. Maneuvering tended to be sedate in all axes, the rudder inducing roll only gradually, with effort needed to counteract the nose's inclination to drop during turns. However, the trim wheel (on the central console) imparted precise control, even for those accustomed to the powered pitch trim of more modern types.

A sticker-shaker – truly innovative in early Viscount days – warned of an impending stall by vigorously shuddering the control column. The device depended on a 0.5 sq. in. spring-loaded vane, heated to prevent icing, projecting from under the port outer wing. In normal flight, air pressure overcame the spring tension, keeping apart a pair of contacts. At 8-12 kt above stalling speed, reduced pressure allowed the spring to close the switch contacts and operate the stick-shaker motor. Another well-regarded advance was a master alarm system that flashed a single red light in front of each pilot whenever any of 20 system control warning lights illuminated.

Early V.700s and some later models carried a radio officer or, in ill-provided regions, a navigator. Otherwise, radios were controlled from the right hand seat or ceiling panel. Navigation tools followed developing technology. BEA fitted Decca Navigator equipment, which drew signals from chains of three transmitters for position fixes; the associated Decca Log recorded progress as a pen-trace on a moving map display. Later, the airline adopted VOR/DME apparatus then gaining currency worldwide. An innovation was the Zero Reader, a predecessor of the flight director, which merged heading with roll and pitch signals from the horizon gyro unit. Progressively improved versions of Smiths automatic pilot were incorporated, although TCA installed Collins autopilot systems in mid 1957. Whilst not standard in any variant, weather radar furnished many Viscounts on delivery; over the decades, a variety of radar equipment found a home as a retrofit in almost all others.

Flaps could be set to 20 degrees for standard takeoff, 32 degrees for maximum lift and minimum drag, 40 degrees for approach or overshoot, or 47 degrees for final approach and landing. Some operators referred to flap settings as, respectively, 43, 68, 85, and 100 percent. On the cockpit pedestal, the flap control was linked with the throttles such that the flaps could not be lowered fully with any throttle open beyond one third. As vitally, if any throttle did open beyond this position with flaps in landing configuration – as when beginning an overshoot – the flaps automatically withdrew to the optimum climb-out setting. (If necessary, the pilot could forcibly override this feature.)

The cockpit of an early "Americanized" V.700, with the fire control switches mounted on the coaming ahead of the pilots. Nosewheel steering was available to either pilot. This aircraft has not yet been equipped with radar. (Brooklands Museum)

Landing gear extension took about 10 seconds. As confirmation of the "down and locked" green lights, a white rod projected an inch above each wing surface to the rear of the inboard nacelle, with another visible through a window in the cockpit floor. In emergencies, muscle-power and a hydraulic pump in the luggage area aft of the cockpit would perform the task. Most Viscounts were equipped with Maxaret anti-skid brakes.

The Viscount's propellers were non-reversing, but featured an ingenious "ground fine" automatic pitch setting to lend braking, complemented by a "flight fine" pitch to prevent the blades from fining off in flight. Takeoff power forced the propeller blades to coarsen to absorb the increased torque until they passed the one-way pitch stop, which then locked to inhibit the blades from regressing below 21 degrees. (The Dart 525 and 530 incorporated a further stop, at 35 degrees, to cater for the higher cruising speeds of the V.810.) On landing, with power under 14,200 rpm (for the Dart 510; figures differed slightly for other versions), shock absorber compression unlocked the pitch stop via a microswitch, per-

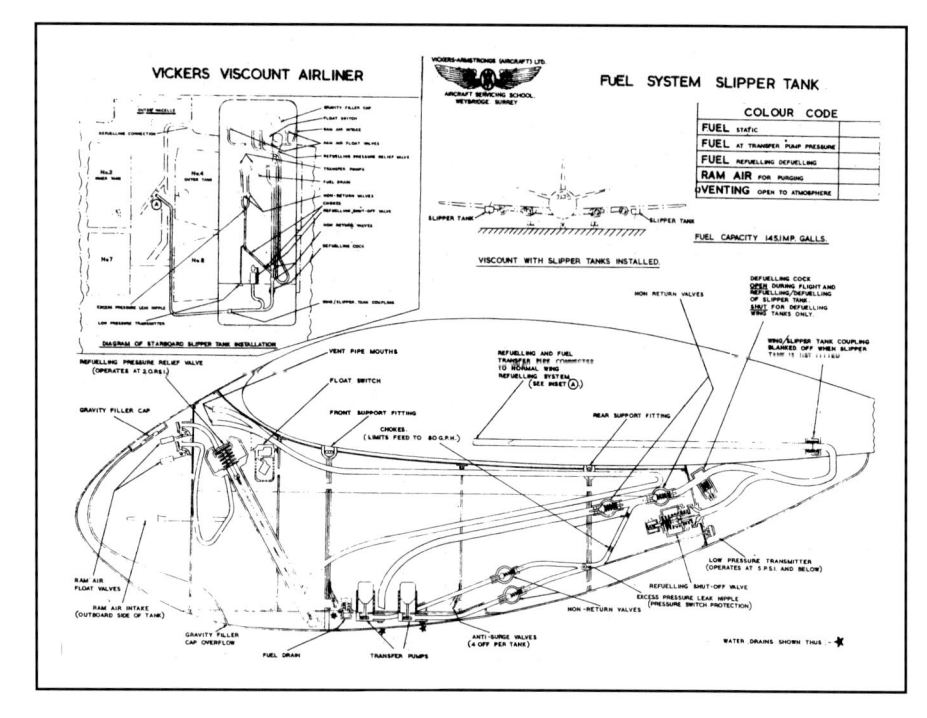

A cutaway view of a slipper tank. The tank was attached to the wing underside at four points, two on the bottom spar boom and two on the leading edge member. With the tank removed, a blanking cap was fitted to the end of the fuel connection and mushroom-headed bolts plugged the fitting holes in the wing skin. (Vickers-Armstrongs via Author)

The open radome of an Ansett-ANA Viscount 832, displaying the Bendix unit fitted to most radar-equipped V.810s. (Brooklands Museum)

mitting the propeller angle now to shift towards zero degrees – ground fine pitch. Because this would occur as soon as the main gear touched the runway, it was preferable to land in a flat attitude, otherwise the abrupt drag would bring the nosewheel down hard.

Since the propeller was directly connected to the turbine driving the compressor, ground fine pitch also ensured efficient use of power on the ground, when less energy was required of the propellers, and eased engine starting by minimizing airscrew resistance. To prevent a propeller from flipping into ground fine pitch following engine failure, microswitches connected to the throttles ensured that, if torque dropped below 50 psi while the power lever – regardless of actual RPM – remained set for takeoff or cruise, feathering would automatically occur.

Feathering an airscrew involved switching the high-pressure fuel cock to "feather," thereby passing through the "off" position and closing the fuel supply, then pushing the feathering button. This energized a hydraulic pump that drove the blades into the fully feathered position (about 83.8 degrees). Operating both switches together could stop the propeller in two seconds. To unfeather and restart an engine in flight, the high pressure cock was shifted to "open" – which automatically directed the hydraulic feathering motor to force the blades away from the fully coarse position – and the feathering button was pulled out; this allowed the propeller to windmill and ignition to occur. The minimum oil temperature for relighting, however, was –30° C, with accelerating the engine beyond idling not permitted below –15° C; in effect, this restricted the operation to warmer – therefore lower – altitudes.

The Problem of Ice

Few Viscount systems brought genuine trouble. But the highly effective hot-air ice protection method introduced *anti*-icing to many crews accustomed to *de*icing. Perhaps unfortunately, when referring to airframe ice protection, manuals generally persisted with the latter term, although it remained valid for engine protection. Anti-icing was mandatory whenever OAT was 10° C or lower, and encouraged if doubt existed.

Sadly, disasters occurred. In April 1958, approaching Tri-City Airport, Michigan, a V.745D of Capital Airlines stalled during a steep turn and struck the ground. Anti-icing had not been switched on, and the horizontal stabilizer had become ice-coated. Tailplane icing similarly caused a Continental Airlines Viscount to dive into the ground while attempting to land at Kansas City one night in early 1963. Although the airfield was above the critical temperature, the aircraft had recently flown unprotected through cloud where measurements were well below 10° C. In 1977, a Skyline Sweden V.838 pitched down and dived into a built-up area 3.5 miles from touchdown at Stockholm (Bromma). The inboard engines – on which the system depended – had been left at low power for so long during the descent that the anti-icing had dropped below its effective temperature. Monitoring the system, in other words, was as important as activating it.

Keeping ice away from the powerplants merited no less alertness. Engine ice ingestion presented special dangers: power loss might be great enough that autofeathering cut in and stopped the engine. Delay in arming engine deicing initiated dis-

A graphic explanation of the use of fuel trimming control, which assisted in ensuring the correct mixture ratio despite variations in temperature, air pressure and fuel density. (Ansett-ANA via Author)

At Right: *Among the most memorable and popular Viscount features were its extraordinarily large windows. This publicity photograph shows an emergency exit; at first, all Viscount windows functioned as exits.* (Brooklands Museum)

Above: *In landing configuration, the Viscount's double-slotted flaps extended to 47 degrees. A V.814 originally with Lufthansa, G-AYOX (c/n 370) flew with British Midland from 1978 until 1984.* (BBA Collection)

VICKERS VISCOUNT

aster in much this way for a Capital Viscount in January 1960, when all engines simultaneously failed. Worse still, loss of the engines allowed electric power to fall too low for unfeathering and engine relight. As recently as 1994, broadly similar circumstances destroyed a V.813 of British World Airways on a cargo flight.

The Viscount Structure: Cause for Concern?

Mindful of the need for low operating costs, Vickers worked hard to minimize weight and keep maintenance simple. Partly also for ease of development, the designers had therefore preferred conservative principles of engineering and manufacture – but had not shrunk from innovation where the rewards justified it. The result was a straightforward design with unfamiliar elements. Most large aircraft possessed a wing built around three spars, with a mid- section. The Viscount's lone wing spar – with "ribs more or less hanging from it like a Christmas tree," as one airline's director of engineering put it – at first invited disquiet.

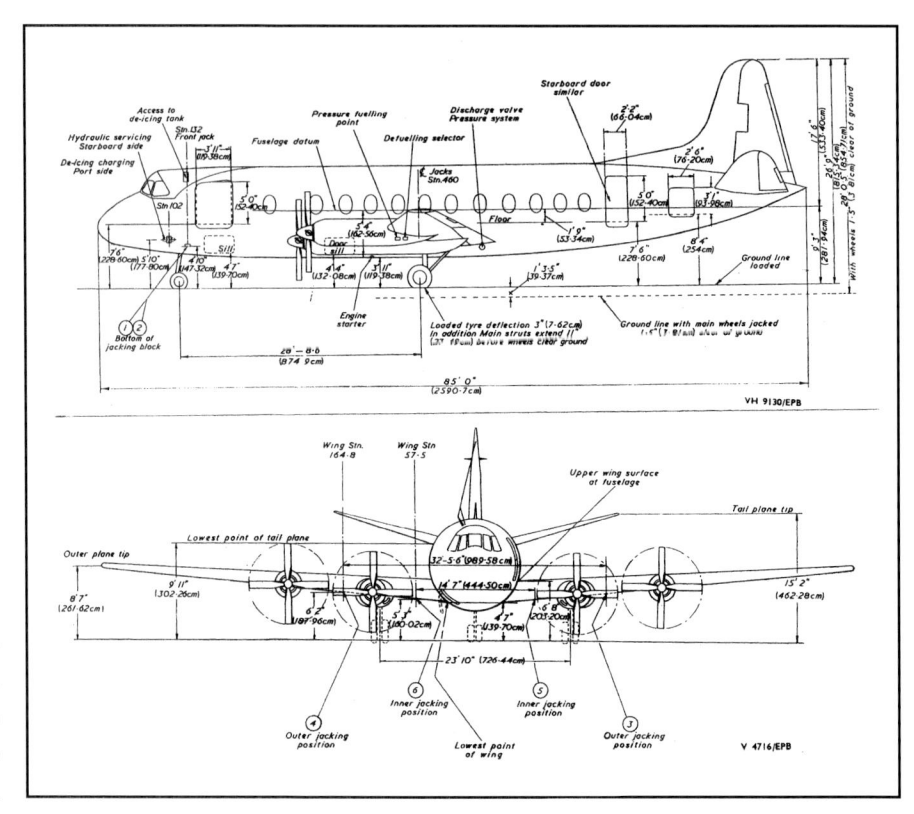

Lateral and front diagrams of the Viscount 800 layout, showing dimensions and clearances. The Viscount 810 was mainly similar, but with a narrower forward door and with a nose radome as standard equipment. (Vickers-Armstrongs via Author)

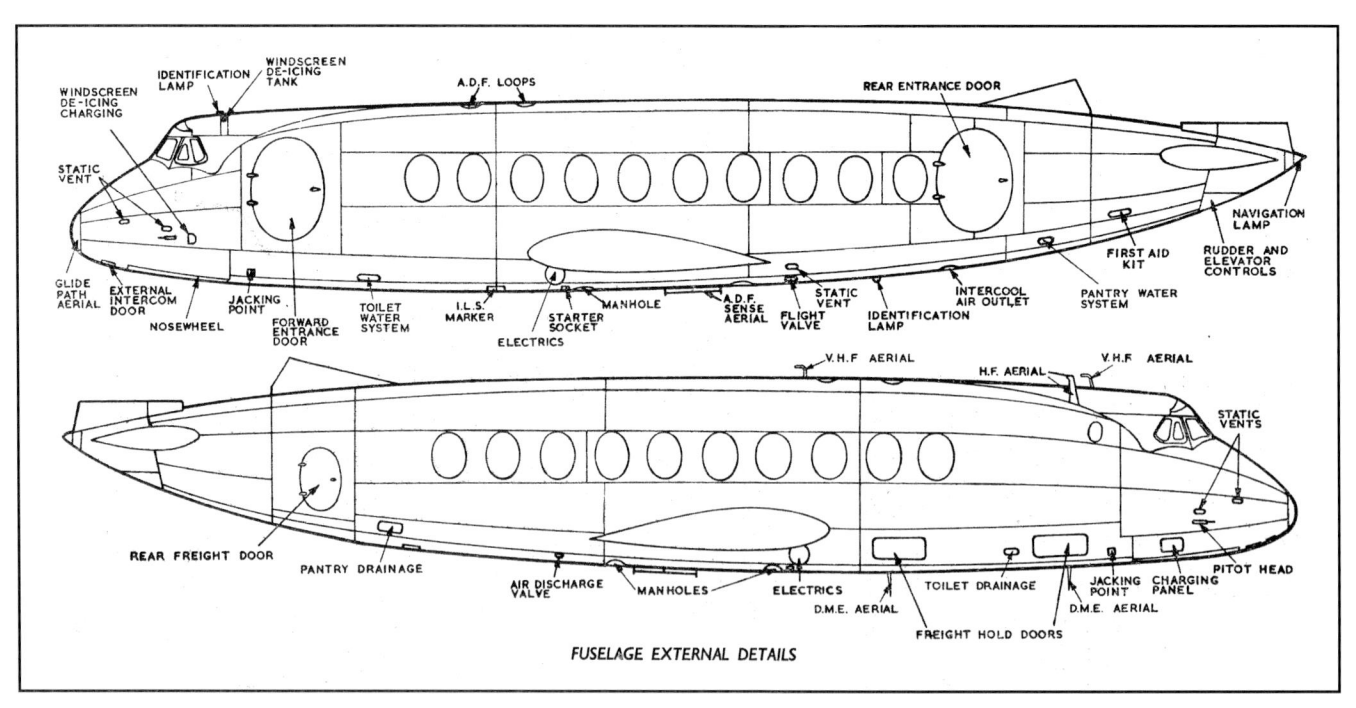

The exterior of the Viscount 700, showing locations of various servicing points and other details. (BBA Collection)

Although the single-spar concept contained nothing inherently amiss, systematic testing in 1953 of two production-standard wing halves did reveal deficiencies. Both trials ended in failure of the lower spar boom in the innermost section. Ensuing calculations indicated spar integrity could not be guaranteed much beyond five years at reasonable utilization levels. Remedies included reducing stress concentration around attachment points, using a tougher, copper-based alloy, and – following evidence gained in testing a third half wing – introducing a plastic strip between the spar boom and the wing skin to eliminate chafing stresses. The spar safe life restriction jumped from 17,000 to 30,000 hours.

Vickers then developed a still stronger spar, incorporated first in Capital's V.745Ds. In 1956, following emergent industry practice, the manufacturer began pressure-testing a complete V.700 airframe in a water tank, and equipped all active Viscounts with strain gauges at critical airframe points, building an invaluable record of structural loads in real-life conditions. But earlier Viscounts were left behind: even 30,000 hours was a mediocre life expectancy, making eventual re-sparring unavoidable. TCA scheduled its Viscounts for spar replacement at its Winnipeg overhaul base during 1958. However, early in that year, it surfaced that original spars might be weakening faster than predicted, triggering a worldwide grounding of 21 unmodified aircraft for immediate rectification.

It was a sensitive moment: in a Viscount crash ten months earlier, metal fatigue had already taken lives. Although the culpable item formed no part of the aircraft's primary structure, the result was no less terrible. A mile from touchdown after an Amsterdam-Manchester journey on 14 March 1957, G-ALWE (c/n 4), the very first production Viscount, suddenly banked into a tightening right-hand turn. The wingtip struck the ground and the aircraft collided with houses. Within days, investigators found that a bolt holding the bottom of the inner starboard flap sections had fractured due to fatigue, causing the flap to jam the aileron control lines. Immediately, all Viscounts were inspected and a redesign instituted to eliminate any recurrence.

Ultimately, Vickers' efforts paid off, evicting doubts and leading to the robust V.810. In truth, structural failure was uncommon; as an accident cause, it was secondary, usually provoked by confrontation with violent weather. Such tragedies included an Ansett-ANA V.720 whose lower spar boom failed in 1961 fol-

On receiving a fleet of ex-BEA V.701s, Cambrian Airways retained the forward airstairs installed by the previous owner. G-AMOG (c/n 7) is now preserved at the RAF Museum at Cosford. (J. Roger Bentley Collection)

The installation of the pantry in the Viscount 802; note the hinge mechanism of the parallel linkage main entrance opening method. (British World Airways via Mike Sessions)

In Viscount 806s engaged on short-haul work later in their career, this unit provided abbreviated snack service. (British World Airways via Mike Sessions)

At London (Heathrow) in mid 1956, F-BGNU (c/n 38) prepares to depart for Paris. Although BEA's Viscounts flew to most other European capitals, until 1957 Air France's V.708s offered the only Paris-London turboprop connection. (VIP Photoservice/Nicky Scherrer)

YI-ACM (c/n 69), a V.735, flew for Iraqi Airways for over 22 years, surviving a disastrous landing at Heathrow in 1955 that necessitated a total rebuild. The airline's Viscounts initially offered a Baghdad-London schedule with all-first-class seating for only 26 passengers. (Carl Ford)

Capital's famous "nighthawk" livery in its element: N7429 (c/n 127) prepares for a night departure from the airline's Washington base. (J. Roger Bentley Collection)

66

Bought new, this V.769D of PLUNA remained with the Uruguayan airline for 24 years. CX-AQN (c/n 321) was photographed in 1972. (Bo Lundkvist via J. Roger Bentley)

Originally a V.745D, CF-DTA (c/n 229), seen at Heathrow, was altered to a V.797D and served for 24 years with the Canadian Department of Transport. The elaborate livery even decorates the slipper tanks. (Brian Stainer via J. Roger Bentley)

Although owned, maintained and crewed by Airwork, this V.831 (c/n 419) was delivered direct to Sudan Airways as ST-AAN. It flew regular 17-hour Khartoum-London services, plus additional Khartoum-Cairo rotations, fitted for 61 passengers. Here, the aircraft lacks its usual slipper tanks. (J. Roger Bentley)

VICKERS VISCOUNT

67

The first turbine airliner delivered in North America was CF-TGI (c/n 40). In 1963, TCA leased it to Transair (Canada), in whose markings it was photographed at Winnipeg. Today, this V.724 is on display at the Pima Air Museum, Tucson, Arizona. (J. Roger Bentley)

N905G (c/n 183), a V.764D, remained in private hands most of its career. Seen here shortly before its October 1977 purchase as HC-BEM by SAN, of Ecuador, it lasted mere weeks in airline operation, crashing that December with the loss of 24 lives. (Author's Collection)

In the 1960s, Aer Lingus, a major Viscount adherent, forsook its green-roofed identity for this more conventional appearance. EI-AKL (c/n 423), photographed at Dublin in April 1965, displays the outsized forward door unique to the V.800 series. (J. Roger Bentley Collection)

G-AOJD (c/n 153) in BEA's "high speed jack" livery: the Air France sticker near the rear entrance indicates a jointly-operated German internal service, for which it and other Viscount 802s were roomily laid out for only 53 passengers. (J. Roger Bentley Collection)

Viscount 798D G-AVED (c/n 286) of BKS Air Transport receiving attention in the hangar; note the mobile office units. In 1970, BKS became Northeast Airlines. (J. Roger Bentley Collection)

South African Airways' Viscount 813s initially accommodated 52, but most became 66-seaters for low-fare services and continued to sustain SAA's domestic operations until 1972. In later livery at Durban in March 1969, ZS-CDX (c/n 350) is the focus of much low-tech activity. (J. Roger Bentley Collection)

VICKERS **VISCOUNT**

69

La Urraca operated this V.837 for only eight months in 1972. Leased, with two others, from TAC Colombia, HK-1412 (c/n 440) handled domestic schedules from Bogotá and Villavicencio. (Bo Lundkvist via J. Roger Bentley)

By the time this photograph was taken at Basle-Mulhouse in November 1972, F-BGNO (c/n 16) had been considerably modified since its Air France days, boasting a full-sized nose radome and improved avionics (note the blade antenna). (VIP Photoservice/ Nicky Scherrer)

Unlike most Viscounts, in the 1980s the V.843s of CAAC still boasted attractively thin logbooks. In 1983, Bouraq Indonesia Airlines bought four, including PK-IVZ (c/n 451); seen at Djakarta in March 1992, it became one of the last passenger-carrying Viscounts active on scheduled operations. (Paul J. Hooper)

Trans-Australia Airlines, an early Viscount operator, later added V.816s equipped with airstairs, including VH-TVQ (c/n 434), seen at Melbourne (Essendon) in August 1968. (J. Roger Bentley Collection)

G-BFZL (c/n 435) spent some of its later years with British Air Ferries; after BAF became British World Airlines in 1993, this was the only Viscount fully repainted. (J. Roger Bentley Collection)

Ghana's West Africa Air Cargo bought this V.814 from Alidair in September 1977. Unfortunately, 9G-ACL (c/n 342) lasted no further than a landing accident in Liberia the following June. The extended double-slotted flaps are clearly displayed. (Carl Ford)

In 1971, BEA devolved all UK regional operations not handled by Cambrian and Northeast to two new subsidiaries – Scottish Airways and Channel Islands Airways – operating transferred V.802s carrying a combined identity, as seen on G-AOHI (c/n 152). Announcement of the British Airways multi-merger soon thwarted plans for separate, permanent liveries. (Carl Ford)

At Gatwick in BUA's late-1960s livery, G-APTD (c/n 426) arrives in classic style: after landing, crews customarily shut down two powerplants to save fuel while taxiing. The V.833 was the only model delivered with the 2,030-ehp Dart 530, although some other V.810s were retrofitted. (J. Roger Bentley Collection)

One of the final Viscounts in service with British Airways was G-APIM (c/n 412), a V.806 originally delivered to BEA. Pictured in 1979, the aircraft later joined BAF, in whose insignia it is now preserved at the Brooklands Museum. (J. Roger Bentley Collection)

Viscount 6 at Work

Worldwide Service; Worldwide Success

The chronicle of British European Airways owes more to the Viscount than to any other airliner. The aircraft's ability to operate fully laden over longer sectors and fly further each day dramatically refashioned the airline's fortunes. In their first year, BEA's Viscount 701s flew more than 6.5 million miles and transported over 335,000 passengers, pushing the carrier's share of air traffic between Britain and continental Europe above 50 percent. In the following fiscal year, V.701s served all BEA's trunk destinations and collectively made a net profit of nearly £1 million – immense for the times.

BEA was not alone. Air France, until recently burdened with obsolete equipment such as the Sud-Est Languedoc, found its profitability astonishingly transformed. For Aer Lingus, the Viscount sparked an extraordinary rise in traffic, with initial load factors averaging 85.5 percent – mirroring the experience of Trans-Australia Airlines. The Viscount's leap ahead of piston airliners in reliability was stunning: during 1955–56, the four Aer Lingus V.707s attained 99.5 percent regularity.

TAA's V.720s entered service in December 1954, and averaged 8.96-hr. daily utilization during their first full year. TAA later purchased seven V.756Ds, and undertook undercarriage modifications – adding automatic braking and reducing tyre pressure from 110 to 85 psi – to enable service to secondary airports, after runway damage problems at some small airfields affected Butler Air Transport. Majority-owned by ANA, Butler became a Viscount operator in autumn 1955, and for a time delivered the world's highest Viscount utilization rate – 9.13-hr. daily.

The Viscount also swiftly infiltrated the network of Trans-Canada Air Lines. The inaugural TCA Viscount flight – the first turbine-powered commercial air service in North America – left Montreal on 4 April 1955, for Toronto, Fort William and Winnipeg. Montreal-New York rotations began three days later; by November, services traversed the continent. In the first year, average load factor exceeded 80 percent, while TCA's market share on the Toronto and Montreal routes to New York jumped by one third; moreover, costs had settled at seven percent below TCA's original projections.

Domestic turboprop travel reached the USA on 26 July 1955 when a Capital Airlines V.744 left Washington for Chicago. On that crucial route, the Viscount's extraordinary passenger appeal catapulted Capital past the competition within weeks, returning load factors of 90 percent and doubling the airline's market share. Starting with Pittsburgh, Cleveland, and New York, the Viscount spread through Capital's system. Frequencies rose as the fleet grew, while night services boosted utilization. Numerous smaller airports, too, welcomed the Viscount: by February 1957, when the last aircraft of 60 arrived – six weeks ahead of contract – Viscounts roamed almost the entire eastern USA, and Capital had ordered 15 more.

At first, the Viscount remained a niche airliner. Beyond TCA, BEA and Capital, few world-level carriers chose it to modernize their medium-range strategy. An exception was Air France, whose dozen V.708s shouldered all the airline's intra-European services for several years from the mid 1950s. Otherwise, the Viscount mainly attracted smaller airlines

An early customer for the V.800 series was the British independent, Eagle Airways, which bought two new V.805s, including G-APDX (c/n 312), to support a move into scheduled operations. The livery featured a red cabin roof. (BBA Collection)

VICKERS VISCOUNT

CF-TGX (c/n 142) of TCA receiving attention in May 1959. The aircraft is a V.757, built to V.700D standards but with the lower-power Darts of the airline's earlier V.724s. (Robert Griggs via J. Roger Bentley)

looking to build regional systems. Trinidad-based British West Indian Airways, for example, experienced dramatic traffic increases, using four V.702s that eventually handled a Caribbean network stretching from New York to British Guiana. Later, V.772s replaced the V.702s, which all found their way to Bahamas Airways to cover similar routes. These aircraft were owned by BOAC Associated Companies Ltd., organized to acquire equipment for BOAC-related airlines and which played an important part in establishing the Viscount among far-flung operators. For Middle East Airlines, for instance, BOAC secured Viscount 754Ds that enabled MEA's traffic growth in 1957–58 to reach over 100 percent, with utilization as high as 9.5-hr./day.

In establishing the Viscount as an airliner as at home in the tropics as in the Canadian winter, the BWIA affiliation was significant. Linea Aeropostal Venezolana soon followed, from March 1956 operating V.749s over an extensive regional and domestic structure. Central African Airways – joint airline of Northern Rhodesia, Southern Rhodesia and Nyasaland – awarded the Viscount its most rigorous task yet. The carrier's V.748Ds routinely took off from hot-and-high airports and flew further than customary for a medium range system. Fitted with specially developed de Havilland airscrews, they were also the first Viscounts delivered with weather radar. To a network ranging across central and southern Africa, CAA added a remarkable Salisbury-London schedule. On this 26-hour expedition, outfitted for 47 passengers and using slipper tanks, the Viscounts commonly took off at 63,000 lb.

Plainly, the Viscount was offering not only a step up from the DC-3, but also proving viable on operations calling for aircraft in the DC-6 class. Iraqi Airways, for instance, worked its Viscounts from 1956 over a regional and internal network, but also from Baghdad to London. Similarly, Misrair introduced three V.739s early the same year and used them not only within North Africa and the Near East, but also to reach Western European destinations from Cairo.

Such versatility would be useful to the wide-ranging needs of Britain's independent airlines. From 1958, Hunting-Clan Air Transport and Airwork employed Viscount 700s on shared schedules (not unlike CAA's operation) from the UK to East and West Africa. The two airlines had originally ordered Viscounts in 1955–56 for military charters, but evaporation of the contracts compelled both to dispose of the new aircraft: Hunting's V.732s spent time leased to MEA, while Airwork sold its three V.755Ds to Cubana, on whose Havana-Miami route they doubled traffic within three months.

When Hunting-Clan again found itself with freshly-built but redundant Viscounts, Icelandair snapped them up. From mid 1957, the two V.759Ds flew domestic and international sectors. Operations could be taxing: winter weather might simultaneously close all Iceland's airports, leaving no alternates nearer than Scotland or Norway. The solution, although it diminished payload, lay in seasonally fitting slipper tanks to guarantee range.

A Longer Fuselage: The Viscount World Expands

The Viscount took a major step forward in 1957, when the stretched variant made its debut. BEA led the way, launching V.802 service that February. The all-tourist, 57-seaters became familiar at Paris, Amsterdam and similar cities, and undertook night freight flights. Aer Lingus' network, with every sector less than 400 miles, made ideal territory for the new model: 60-seat V.808s succeeded its V.707s in 1957 on routes such as those to London and Amsterdam. The higher-powered V.806, configured for 16 first class and 42 tourist passengers, joined BEA in January 1958. Sixteen months later – by which time Viscounts maintained almost the entire BEA timetable – a V.806 opened the first scheduled London-Moscow flights by a British airline.

The Viscount 800 proved to be an interim variant. Nevertheless, it earned Vickers new customers among carriers needing a short range, high-density workhorse. In June 1957, KLM placed V.803s on its shorter European sectors. New Zealand's domestic carrier, National Airways Corporation, began V.807 service in February 1958, doubling traffic on main routes within two years. Such was the type's success that NZNAC dropped a scheme to reduce seating and expand freight accommodation, actually adding seats instead. In Britain, Eagle Airways purchased two V.805s for scheduled services connecting several British cities with continental points. The pair individually flew

Stubborn competition: the Convair 240, 340 (pictured) and 440 captured sales to KLM, Iberia, TAA, Sabena and Lufthansa, among others—including Alitalia, which acquired Viscounts only by amalgamating with LAI. Nevertheless, many later switched to Viscounts. (Author's Collection)

Norway's Fred. Olsen specialized in charters and leasing. Delivered in 1957, its V.779Ds wore this livery comparatively little, instead spending over two years with Austrian Airlines. LN-FOK (c/n 252) also flew for SAS before being sold in 1962. (ATPH Transport Photos)

TAA supplemented its short-body Viscounts with two V.816s in 1959, including VH-TVQ (c/n 434). By April 1960, the world's highest-time Viscount was a TAA V.720. (ATPH Transport Photos)

also for Eagle's Bermuda subsidiary on strikingly successful Bermuda-New York schedules.

Meanwhile, the Viscount 700D was far from eclipsed. Italy's postwar intercontinental airline, Linee Aeree Italiane, introduced Viscount 785Ds on European schedules in spring 1957. Within months, LAI amalgamated with Alitalia, in whose markings the final four of 10 were delivered. After jet services opened, Alitalia not only retained Viscounts to operate lighter international routes and replace smaller types on domestic sectors, but even bought eight used examples. In May 1957, Philippine Airlines accepted its first V.784D; on the Manila-Hong Kong link, PAL's Viscounts boasted load factors close to 90 percent within a month. Union of Burma Airways employed V.761Ds from 1957 on regional flights as far as Singapore. Also in 1957, via BOAC Associated Companies, two Viscount 760Ds joined Hong Kong Airways, and reliably connected Hong Kong with points including Bangkok, Manila and Tokyo. In 1959, they passed to Malayan Airways, whose regional routes included a busy Kuala Lumpur-Singapore link. Subsequently, the same pair flew for Aden Airways, ranging as far as Khartoum and Nairobi.

Repeating a common theme, the introduction of Viscounts by Indian Airlines in October 1957 instigated remarkable traffic increases. The turboprop's qualities helped offset India's commercially perilous combination of expensive fuel and low fares. Viscounts visited every major Indian city, plus regional destinations abroad, maintaining a high

Airwork's V.831's were highly unusual among V.810s in retaining the wider V.800 forward door, as seen on G-APND (c/n 402). (BBA Collection)

work-rate until reassigned in 1962-63 to shorter sectors and night-mail operations. Iranian Airlines confined its V.782Ds, received in 1958, to a regional network. However, in Turkey, five V.794Ds delivered via BOAC to Türk Hava Yöllari would eventually take Western European voyages in their stride.

For TCA and Capital – although neither bought stretched versions – Viscount activity was reaching a peak. By 1958, TCA's Viscounts flew 10 daily New York rotations from both Montreal and Toronto, and were ubiquitous in Canada and at important airports in the USA. After Vanguards arrived in 1961, the dependable Viscounts increasingly plied shorter routes connecting lesser points. Matters went less smoothly for Capital. Locked within a high-cost, short-sector network, the airline sought profitability in vain, and had trouble making payments on so many new airliners. Competitors added pressure with new, faster aircraft on prime routes, and, by 1960, debts were surging. Owed $33.8 million, Vickers foreclosed. United Air Lines rapidly proposed a takeover, and the resulting deal, finalized in June 1961, included 41 of the surviving Viscounts. Vickers took back the 15 others – but, within a year, a surprised United found the type so productive that it re-purchased six of them.

Capital's tribulations had forced it to retract its repeat order for 15 Viscounts. Northeast Airlines snatched up eight of the completed aircraft in 1958, adding two of its own. From Boston, Northeast's V.798Ds built up business on the high-traffic New York run – on which daily rotations reached 21 by 1960 – also providing lively regional connections. Like Capital, however, Northeast grew financially overstretched. The challenge of Eastern

Note the "radar equipped" legend on the nose of Capital's N7457 (c/n 212) at Chicago Midway in September 1959. (VIP Photoservice/Nicky Scherrer)

Airlines' no-reservations shuttle on the vital Boston-New York route proved fatal, and, as Northeast collapsed, Vickers repossessed the aircraft in 1963.

The Viscount in its Prime: The V.810 Arrives

The advent of the radically enhanced V.810 series boosted the Viscount to its zenith in airline appeal. First deliveries went to Continental Airlines, whose 15 V.812s boasted weather radar, airstairs and a four-place rear lounge. The inaugural service left Chicago for Los Angeles on 27 May 1958, staging through Denver, the carrier's base. The Viscount network ramified, especially through the Southwestern United States, where the type's superior hot-and-high abilities proved advantageous. Despite the

G-APTA (c/n 71), a V.702, joined Kuwait Airways in 1959-61 on lease from BOAC Associated Companies. Many early Viscounts like this one faced permanent retirement in the 1960s due to their lack of later wing spar enhancements and ineligibility for US certification. (Author's Collection)

VICKERS VISCOUNT

As the 1960s arrived, early Viscounts were finding work among British independents such as Tradair. G-APZB (c/n 30) came from Aer Lingus after the Irish airline had decided against converting its V.707s to freighters. (ATPH Transport Photos)

breakthrough suggested by the early Continental purchase, however, the upgraded Viscount faced indigenous competition from the Lockheed Electra. California Eastern Airlines did order eight V.823s in 1957, but had to cancel when expected route awards evaporated. Continental's remained the only airline V.810 sale in the USA.

After becoming a Viscount airline by merging with LAI, Alitalia enthusiastically sought further examples. Seen at Heathrow in 1964, I-LIRC (c/n 114) was one of five V.745Ds acquired from the batch returned to Vickers after United's takeover of Capital. (ATPH Transport Photos)

The V.810 nevertheless briskly won orders not only from the world's regional airlines, now eager to replace timeworn piston aircraft with more efficient types, but also from prominent flag carriers such as Lufthansa. The German airline began Viscount flights to main European points in October 1958, followed by Middle Eastern services. South African Airways introduced V.813s in late 1958, rapidly transforming its hitherto loss-making domestic network. Pakistan International Airlines enlisted the first of its five mixed-class Viscount 815s in January 1959. Although two perished in early accidents, PIA employed the survivors for a decade on Pakistan-India connections and on services stretching from the Middle to Far East. Meanwhile, the first Viscount 827s for VASP reached Brazil in autumn 1958 and soon permeated the VASP system, with load factors rising as high as 99 percent.

Elsewhere in Latin America, progress was erratic. When stalled financing defeated the Viscount plans of Lloyd Aereo Colombiano, two of its three ordered V.786Ds passed to LANICA in early 1958. Based in Nicaragua, LANICA's timetable featured a through service from Miami to Lima, Peru. The carrier sold both aircraft a year later, one to TACA of El Salvador, who had served major regional points with a leased V.784D since late 1957, and had purchased the third redundant LAC machine. In Uruguay, PLUNA brought its first Viscount 769D into operation in mid 1958; supply of two others waited 12-18 months for financing. Eventually, a fleet expanded by secondhand aircraft would connect Montevideo reliably with main destinations for almost 25 years. Cubana, too, suffered financing problems, delaying its ordered V.818s until December

1958. Within weeks, Cuba's revolution presaged the end of direct Cuba-US air links, robbing the airliners of their best routes.

Both main Australian domestic airlines supplemented their short-body Viscounts with new variants. Having acquired V.747s by absorbing Butler Air Transport in 1958, a year later Ansett-ANA received four V.832s – the first Viscounts certificated at 72,500 lb. AUW, affording 25 percent more payload over extreme ranges. Swelling traffic prompted rival TAA to add two Viscount 816s in mid-1959. Both TAA and Ansett-ANA also bought secondhand V.810s. In 1962, an Australian government attempt to engineer parity between the two airlines by mandating an equipment swap saw three TAA V.720s leased to Ansett. Ansett also bought a further TAA V.720 outright.

Small V.810 orders came from UK independent airlines. Unfortunately, a lost military contract forced Eagle Aviation (affiliated to Eagle Airways) to cancel, but Airwork and Hunting-Clan received several in 1959 for African routes. The V.833s of Hunting-Clan used higher-powered Dart 530 engines, later retrofitted to Airwork's V.831s. Supplied with nose-mounted, ground-mapping radar for their role in poorly equipped regions, the aircraft usually carried slipper tanks. The Airwork/Hunting-Clan alliance climaxed in the 1960 creation of British United Airways, whereupon the V.810s soon forsook most African duties.

Seen after returning from a spell with Eagle Airways' Bermuda subsidiary, VR-BBH (c/n 32) is about to become G-ARKI with the newly-renamed Cunard Eagle Airways. The small starboard porthole aft of the cockpit was characteristic of most variants. (J. Roger Bentley Collection)

One of the ex-Capital V.745Ds sold back to Vickers by United Airlines was N7420 (c/n 118), seen in residual Capital livery and temporarily registered G-ARHY. It became PI-C773 with PAL. (BBA Collection)

VASP acquired a number of BEA's V.701s in 1962–63, removing the airstairs previously retrofitted. PP-SRS (c/n 182) awaits delivery at Heathrow in August 1962. (J. Roger Bentley Collection)

VICKERS VISCOUNT

On United's absorption of Capital in 1961, N7408 (c/n 106) was one of the earliest aircraft repainted. UAL returned some Viscounts to Vickers, but later repurchased many of them. (Author's Collection)

By April 1960, the worldwide Viscount fleet had completed 2,680,000 hrs. and 1,715,800 landings; 113 airframes had exceeded 10,000-hr., headed by the 14,236 hrs. logged by a TAA V.720. In small numbers, the Viscount still attracted new customers: in anticipation, Vickers laid down a group of airframes temporarily labelled V.810U. Some became V.837s for Austrian Airlines in 1960. (The airline actually began activity in early 1958 using three Viscount 779Ds supplied by Fred Olsen, who then held a 15% shareholding.) From the same batch, Ghana Airways received three V.838s in 1961.

In Japan, the Viscount contributed crucially to the expansion domestic air travel. All Nippon Airways introduced Viscounts on principal routes from Tokyo in mid 1961. Immediately popular, ANA's V.828s featured television sets fitted on the overhead racks, making them probably the first airliners thus equipped. With Viscounts soon trekking 30 times daily each way between Tokyo and Osaka alone, besides fulfilling many other schedules, demand rapidly inflated well beyond the type's ability to cope.

In December 1961, the Civil Aviation Administration of China placed the final Viscount order. The historic contract for six aircraft marked the first from the People's Republic of China for new, Western-

In 1962, BUA sold three Viscount 804s to LOT to supplement the Polish airline's Il-18s. By the time SP-LVC (c/n 248) was photographed at Zürich in March 1966, two had been destroyed. (VIP Photoservice/Nicky Scherrer)

built equipment. A U.S. ban on technology supply to China meant all equipment in CAAC's V.843s normally of US origin – electrical, navigational and hydraulic items, plus instrumentation – was British instead. After delivery between July 1963 and April 1964, China's Viscounts concentrated on domestic requirements and were rarely seen further afield, except in later years on overhaul at Hong Kong.

Years of Maturity: The Viscount as Second-Level Airliner

As the final new Viscount left the factory early in 1964, the story entered its longest and most diverse chapter. Major airlines began to disperse their collections – but, for BEA, years of Viscount activity remained. Many V.701s passed indirectly in 1963–66 to Cambrian Airways and BKS Air Transport (later renamed Northeast Airlines). Both carriers became part of BEA's British Air Services subsidiary, and, from 1968, received most of the parent's V.806s. BEA's Viscounts remained popular, meanwhile, on German internal services, refitted with a roomy, 53-passenger, four-abreast cabin. After V.802s adopted almost all Scottish schedules in 1967, BEA created Scottish Airways and Channel Islands Airways in 1971 by spinning off the relevant activities, dividing its V.802 complement between the two. However, the amalgamation that launched British

KLM purchased nine high-density, short-range V.803s. Seen landing at Frankfurt in 1964 is PH-VIA (c/n 172) in the carrier's final Viscount markings. (VIP Photoservice/Nicky Scherrer)

This view of Continental's N252V (c/n 364) in later livery displays the additional small lounge window and lack of baggage hatch. The V.812 also featured individual emergency oxygen masks – then still a novelty – plus integral airstairs, uncommon in V.810 variants. (J. Roger Bentley Collection)

Only two of Northeast's ten Viscount 798Ds were built as such, including N6598C (c/n 392); the others were originally destined for Capital as V.745Ds. Northeast's clientèle typically were business travellers with light baggage, more than compensating for the weight of the integral airstairs. (Author's collection)

Airways in 1973 reunited all the remaining Viscounts of BEA's four offshoots into a 37-strong fleet, maintaining the same network. A progressive run-down led to final operations in 1981–82, with a small mustering of V.806s principally working Scottish routes.

As TCA became Air Canada in 1964, its numerous Viscounts remained busy. Even on short sectors, traffic gradually outgrew the type, and the 1970s found it confined to Air Canada's eastern network. By spring 1974, the once-dominant armada had been withdrawn. Most met the breaker; a handful found private ownership or a career in Zaire. United's Viscount utilization proved no less frenetic than Capital's: in late 1964, the worldwide highest-time and highest-cycle Viscounts both flew with UAL. Inevitably, aircraft ran short of hours, and United's final Viscount landed in January 1969. Some examples moved to private or minor operators, others provided spares or scrap metal. Several other Viscounts joined Aloha Airlines, flying intensive Hawaiian inter-island sectors until 1971. Rival Hawaiian Airlines employed a V.798D and a leased V.745D during 1963–64.

BEA's V.802 G-AOHK (c/n 160), landing at Heathrow. In the late 1960s, BEA modified the capacity of most V.802s to 66— some, with catering facilities removed, to 71. The airline was also required to enlarge its registration lettering, as seen here. (ATPH Transport Photos)

VASP secured 10 BEA V.701s in 1963, flying them exhaustively alongside its higher-performance models for almost six years until fatigue concerns grounded them. In 1961-63, Air France leased two V.708s to Air Vietnam, but eventually its whole fleet migrated, directly or via intermediate owners, to Air Inter. The French domestic carrier added four V.724s from Air Canada in 1964. Air Inter's Viscounts dominated an industrious, short-sector network; at retirement in the early 1970s, they included some of the highest-time airframes.

Among UK independent airlines, BUA accumulated 14 assorted Viscounts, keeping them busy throughout the 1960s on schedules and inclusive tour work. With used V.755Ds and leased V.701s, British Eagle (descendant of Eagle Airways, via Cunard Eagle Airways) expanded scheduled Viscount operations, and met non-scheduled growth by

For a time, several BEA V.802s, including G-AOHG (c/n 156), wore a different response to the tightened UK regulations concerning registration markings, with large lettering on the fuselage rather than the lower fin. This aircraft also retains the original, barely-visible lettering at the fin-tip. (BBA Collection)

In 1963, Channel Airways acquired G-ALWF (c/n 5) from BEA, along with a number of other V.701s; although BEA had already remodelled them from 47- to 63-seaters, Channel managed to raise accommodation to an astonishing 71. In 2003, this remained the oldest existing Viscount. (J. Roger Bentley Collection)

Another of the undelivered Capital order, PI-C-772 (c/n 227) was converted to Philippine Air Lines specification as a V.784D. The aircraft is seen during servicing in the UK by Marshall's of Cambridge. Note the slipper tanks. (ATPH Transport Photos)

Cambrian Airways, of Wales, changed its paint scheme several times in only a few years. G-AMOO (c/n 28), a V.701, displays the slightly sharpened nose profile of the radar modification earlier applied by BEA. (BBA Collection)

Almost no photographs exist of LN-SUN (c/n 141) in Braathens SAFE markings. Denied approval to serve the routes for which the single V.742D had been ordered, the Norwegian airline sold the aircraft before delivery. (The A. J. Jackson Collection)

adding used V.732s. Channel Airways purchased most of Continental's V.812s in 1966, replacing ex-BEA V.701s on scheduled and non-scheduled passenger and light cargo services. Noted for cramming exceptional numbers of passengers into its aircraft, Channel squeezed 71 seats into its V.701s and no fewer than 80 into the V.812s.

In 1966, Aer Lingus purchased all of KLM's V.803s, and converted three of its V.808s for passenger/freight use. Luxair employed a single ex-PIA aircraft in 1966–69, and various V.810s reached Israel, where Arkia used them throughout the 1970s. Worldwide, airlines also operating handed-down V.810s in this period included Taiwan's Far Eastern Air Transport, TAC Colombia, Lineas Aereas de Urraca and SAN Ecuador. Condor Flugdienst leased several in the 1960s from its parent, Lufthansa.

Central African Airways separated into three airlines in 1967 – an outcome of international sanctions imposed on Rhodesia. Five Viscounts passed to Air Rhodesia for domestic and South African services; Air Malawi received two. Despite problems with the embargo, especially for spares supply, Air Rhodesia kept aloft, even procuring three further V.700Ds.

By the early 1970s, few original fleets remained. Used Viscounts found work with several European charter companies and put down strong roots among UK independent airlines. But Indonesia, too, emerged as a latter-day Viscount stronghold after three BEA V.806s moved to Mandala Airlines in 1969. Mandala absorbed Seulawah Air Services, developing a vigorous system around a fleet sourced mostly from Far Eastern Air Transport. Merpati Nusantara Airlines received its first Viscounts in 1970 from All Nippon, later acquiring others elsewhere.

Nevertheless, Britain soon hosted more Viscount activity than anywhere else. For a decade and a half, a major adherent was British Midland Airways. BMA's Viscount history dated to a V.736 and two ex-PIA V.815s obtained in 1967. After BMA took over Invicta Airways with its two V.755Ds, its fleet steadily mounted, enlarged greatly by former SAA and Lufthansa aircraft. The airline operated both scheduled services and pure charters – including a renowned wet-lease business.

Dan-Air, meanwhile, used a variety of Viscounts on charter and scheduled duties. Several came on lease from Alidair, an operator that started out with ex-Channel Airways V.812s. Alidair offered both passenger and freight charters, and fulfilled oil industry contracts for many years with various early-build V.700s based at Aberdeen. When the airline (renamed Inter City Airways) ceased trading in 1983, these elderly machines passed to Janus Airways for coach-air services linking England and France.

The Sunset Years: The Viscount at Century's End

For British Midland, the Viscount's reliability and depreciated costs made it a profitable workhorse through the 1970s and most of the 1980s. Other UK carriers active into the 1980s included Intra Airways (later Jersey European Airways), Guernsey Airlines, London European Airways and Euroair. In the USA, Viscounts operating as airliners in any traditional sense mostly belonged to travel clubs, who found them economically ideal for their low-intensity needs.

By 1980, Viscounts the world over had amassed 12 million flying hours and 10 million landings, and even secondary operators had now

The final Viscount constructed at Hurn was this V.839 purchased in 1961 for VIP use by the Shah of Iran. EP-MRS (c/n 436) soon passed to the national airline instead. In this 1964 view, note the two blanked off leading windows. (ATPH Transport Photos)

The Viscount's success brought ironic consequences: protracted delivery times deterred some important customers. Swissair almost selected the V.800, but chose Convair-liners, as did SAS. However, a capacity shortage in 1960–61 compelled SAS to lease V.779Ds from Fred Olsen—hence the unfamiliar sight of LN-FOH (c/n 250) in these markings. (Brooklands Museum)

One of only three Viscounts ever to carry Spanish markings, this V.831, EC-AZK (c/n 419), spent part of 1965 leased by Aviaco from BUA. (Author's Collection)

Cyprus Airways used two BEA V.806s on regional services in 1965–70. Despite full livery, they remained under UK registry. G-AOYJ (c/n 259) is about to depart the Beirut ramp in August 1966. BEA also wet-leased V.802s or V.806s to Malta Airways, TAP (Portugal) and Gibraltar Airways. (VIP Photoservice/Nicky Scherrer)

mostly espoused newer types. Airframe longevity was a natural concern: most of the active population had put in well over two decades of toil.

The British Airways aircraft had led lives as hard-working as any; remarkably, many now tackled a fresh career. British Air Ferries adopted the Viscount in 1981, swiftly collecting ex-BA V.802s and V.806s for use as 76-seaters on inclusive tour contracts, separately obtaining a V.815 to handle longer sectors. In 1983, BAF took over Alidair's oil support contract and began basing Viscounts at Aberdeen. Growing scheduled and charter business pushed BAF to expand; by 1984, 21 Viscounts were on strength, including a pure cargo variant. Virgin Atlantic Airways wet-leased a series

Before joining LAV, YV-C-AMB (c/n 24) served BEA as G-AMOK and suffered severe damage in a 1955 accident. Cockpit modifications incorporated during the subsequent complete rebuild earned it the unique designation V.701X. Aeropostal removed the forward airstairs installed by BEA. (J. Roger Bentley Collection)

N820BK (c/n 233) was one of many V.700Ds transformed from airliners into executive transports, in this case for the Blaw Knox Corporation. As here, most retained airstairs. (J. Roger Bentley Collection)

The former Capital network continued as the principal territory of United's Viscounts after the merger. N7448 (c/n 204) was photographed at Washington, DC, in 1966. (J. Roger Bentley)

One of the few tropical airlines not to choose the higher-performance V.700D was BWIA, which replaced several early Viscounts with V.772s, including 9Y-TBT (c/n 236), seen in later insignia. (J. Roger Bentley Collection)

of BAF Viscounts during the 1980s to connect its trans-Atlantic flights with Maastricht and Dublin. Collaborating with British Aerospace (successor to Vickers-Armstrongs), BAF devised an airframe life extension program during the 1980s to permit the aircraft to carry on up to 75,000 flights, reducing the operating ceiling to 17,500 ft. and the cabin pressure differential to 3.5 psi. Later, BAF's Viscounts began operating night package links in 1992 for Parcel Force. In the following year, BAF became British World Airlines, and operated the final UK passenger Viscount service on 18 April 1996.

With the 1980 political resolution, Air Rhodesia became (briefly) Air Zimbabwe-Rhodesia, then Air Zimbabwe. In due course, the airline

Like many Viscount airlines, Ansett-ANA not only purchased new aircraft but also added used examples, such as the ex-Continental V.812 VH-RMK (c/n 355), seen at Melbourne after withdrawal in 1969. (VIP Photoservice/Nicky Scherrer)

For safety, propeller braking operated in conjunction with the built-in forward airstairs. Several Viscounts joined Aloha Airlines in the 1960s, including N7416 (c/n 232), a V.798D earlier repossessed by Vickers from Northeast, seen at Honolulu in 1968. (J. Roger Bentley Collection)

added two used V.810s, but, gradually, the Viscounts reached retirement, the last examples leaving in December 1990. In Indonesia, both Mandala's and Merpati's Viscounts served into the 1990s, as did Bouraq's ex-CAAC V.843s, becoming some of the last Viscounts carrying passengers on scheduled operations. Zaire (now the Democratic Republic of the Congo) has been home to many Viscounts over the years, mostly for charter use, with over a dozen carriers. Elsewhere in Africa, Botswana, Gambia, Swaziland and Togo have all hosted Viscounts, including some of the very last operational airframes.

Corporate, Private, and Official Viscounts

Even as the Viscount turned the airline world on its ear, it began to attract commercial and government organizations looking for an efficient private transport. The US Steel Corp. received three V.764Ds in December 1956, with VIP interior, radar, fuel jettisoning and airstairs. A 450-Imp. gal. belly tank and slipper tanks raised total fuel capacity to 2,640 Imp. gal., conferring a still-air range above 2,500 nm, with 10 occupants. The single V.765D delivered shortly afterwards to the Standard Oil Co.

Winner Airways of Taiwan acquired this V.806 from BEA in 1969. Rarely photographed, B-3001 (c/n 268) saw use as a freighter in Malaysia and Vietnam, and is seen at Bangkok in April 1970. (J. Roger Bentley Collection)

Early Viscounts offered opportunities for many airline start-ups during the 1960s and 1970s. Air International operated G-APPX (c/n 71) during 1972. (Carl Ford)

By the 1970s, even late-model Viscounts were leaving major operators. Partly-painted and still registered VH-RMJ, Ansett's V.832 (c/n 417) is being prepared in April 1970 at Essendon, Melbourne, for sale to Far Eastern Air Transport as B-2017. (J. Roger Bentley Collection)

Originally the prototype V.810 before being refurbished in 1960 as a V.827 for VASP, PP-SRH (c/n 316) arrives at Santos Dumont, Rio de Janeiro, in 1970. VASP's V.827s seated 52 and featured the rear lounge. () J. Roger Bentley Collection)

was mainly similar. Both businesses used their Viscounts comprehensively to carry company executives and customers; those of US Steel in particular travelled widely abroad. The stalled Capital Airlines reorder provided a V.776D for the Kuwait Oil Co. in 1958; another became a V.793D with the Royal Bank of Canada. The other corporate Viscounts sold new, both V.810s, also originated in annulled airline orders. In 1959, the Tennessee Gas Transmission Corp. purchased one originally intended for Cubana, and a cancelled TAA aircraft went to the Union Carbide Co. in 1960. The Niarchos Group, of Panama, ordered one V.819 in 1956, but cancelled.

Little created as much excitement, however, as a 1955 contract for one V.763D from the Hughes Tool Co. The company – owned by the irrepressible Howard Hughes – possessed a large slice of Trans World Airlines stock, and the order specified airline configuration. Naturally, this ignited talk of a TWA order. (Vickers would later try, complete with demonstration visit, to tempt TWA with the undelivered Capital batch.) Typically, after much vacillation – the aircraft only made its first flight two years after completion – Howard Hughes' interest drifted, and the Viscount instead went to TACA.

Among government agencies, the Canadian Department of Transport acquired a V.737 in 1955 (initially leasing it to TCA, complete with titles, to assist crew training) and a VIP-configured V.797D in late 1958, adding a used V.724 in 1964. The non-VIP aircraft performed as mobile offices. More rarefied was the civil-registered Viscount 839 bought in 1961 by the Iran Government for luxury use by the Shah, although it passed to Iranian Airlines in early 1963.

Private employment handed second careers to numerous ex-airline Viscounts. The early 1960s saw some remodelled as executive transports for companies such as Blaw Knox and The Victor Comptometer Corp. Most special conversions drew on V.745Ds handpicked from those withdrawn by United in 1968–69. Ready availability and low purchase cost, along with its performance and ability to cope with any American airfield, made the V.700D a particularly popular choice.

Such factors, together with airline-style comfort and speed, also attracted VIP charter and "private airline" operations. The Go Corporation specialized in this business, chartering Viscounts well into the 1980s to sports teams, touring politicians, and music acts such as Jefferson Starship and Alice Cooper. The company ran a large Viscount maintenance facility at Tucson, Arizona, with a vast provision of spares. Private organizations using Viscounts included religious concerns such as the Copeland Evangelistic Association, the Cathedral of Tomorrow choral group and the Oral Roberts Evangelists Association, which kept a Viscount to transport an affiliated college baseball team. From 1968, Ray Charles employed a rejuvenated V.745D, and later a V.835, to convey himself and his musicians. For many years, one ex-LANICA V.786D made an especially luxurious personal transport for the art collector and philanthropist, Marjorie Merriweather Post. Inevitably, the private ownership record is long and often obscure, entailing many resales and

One of the last V.806s delivered to BEA, G-APKF (c/n 396) just before departing for Lao Air Lines in September 1969 to become XW-TDN. (J. Roger Bentley Collection)

Many European charter operators employed handed-down Viscounts in the 1960s and 1970s. V.814 D-ANUR (c/n 342) is seen in operation with Lufthansa's non-scheduled offshoot, Condor Flugdienst. (BBA Collection)

TC-SET (c/n 432) preparing to leave a wet ramp. THY's V.794Ds undertook not only regional services but also ranged from Ankara as far as Amsterdam. (J. Roger Bentley Collection)

Air Commerz operated this V.808C as D-ADAN (c/n 421) in 1970-72. The size and shape of the double cargo door, with the added small window inset, are clear. The freighter conversion also entailed removing the single standard window ahead of the propeller line. On the carrier's bankruptcy, Aer Lingus repossessed the aircraft. (Carl Ford)

involving individuals as well as corporations, plus brokers, leasing agencies, and others.

Military Operators

Few Viscounts appeared in uniform. The Indian Air Force enlisted two in 1955–56; each possessed differing VIP appointments and therefore different type numbers. After over a decade, both passed to Indian Airlines. Similarly, Pakistan bought one V.734 in March 1956 for government use. In 1970, it went to the Chinese Air Force, serving in a similar role until replaced in 1983 by two ex-CAAC V.843s. Also fitted out for VIP tasks were two Viscounts for the Brazilian Air Force, the first of which had been built and fully painted for – but cancelled by – the Norwegian airline Braathens SAFE. Both served with the 1st/2nd Grupo de Transporte Espescial, under the FAB designation VC-90. A single V.781D joined the South African Air Force in June 1958 for VIP assignment with 21 Sqn., later with 44 Sqn. From 1984, it carried the civilian

During 1972-1976, BOAC (latterly as British Airways) leased two Cambrian V.701s to offer feeder services from Edinburgh and Belfast to connect with trans-Atlantic flights at Prestwick, where G-AMOG (c/n 7) was photographed. (BBA Collection)

At Bangkok in June 1970, V.761D XY-ADH (c/n 190) displays Union of Burma Airways titles, unusually, in English. As with many of the world's local and regional carriers, UBA originally turned to the Viscount as a DC-3 replacement. (J. Roger Bentley Collection)

CF-TIF (c/n 386), seen in modified Air Canada livery ready for sale at Winnipeg in 1974, was originally TCA's penultimate V.757. Politically motivated antipathy to "old-fashioned" propliners hastened the retirement of Air Canada's fleet. (J. Roger Bentley)

identity ZS-LPR. In 1991, as the longest-serving Viscount still with the original purchaser under the same name, it was sold in Zaire to ITAB, later joining Bazair. Its long life ended in a mysterious crash in 1997, allegedly shot down by rebels.

In 1961, the Royal Canadian Air Force seemed ready to buy some of the V.745Ds returned to Vickers in the UAL takeover of Capital, but the idea died. Much earlier, Vickers supplied two stillborn proposals for Viscount 700 variants for the Queen's Flight, one a full VIP model and one a staff transport.

Other military Viscounts came second-hand from the civil world. A V.836 and a V.839, both ex-corporate examples, flew with 34 Sqn., Royal Australian Air Force, in 1964–69. After intermediate civilian owners, the same two aircraft joined the Sultanate of Oman Air Force in 1971, partnered by a V.814 and two V.808s. In 1971-72, the Turkish Air Force took over three THY V.794Ds for general transport duty, retaining them into the 1990s.

Although no Viscounts served with Britain's armed forces, several

By the time HK-1061 (c/n 327) of Aerolineas TAO was recorded at Bogotà in January 1977, it had been out of use for two years. The aircraft, a V.785D, first flew with LAI in 1957, and retains elements of Alitalia markings. (VIP Photoservice/Nicky Scherrer)

VP-YNB (c/n 99), seen at Johannesburg in 1972. One of five delivered to Central African Airways in mid-1956, this V.748D flew with CAA and its successors, Air Rhodesia and Air Zimbabwe, for close to 35 years. The aircraft appears to be in mid-bounce: note the fully-extended nose oleo. (VIP Photoservice/Nicky Scherrer)

One of many Viscounts offered by the Go Corporation for hire as personal airliners, N24V (c/n 228) bears the logo of the rock band Bad Company during a US tour in the mid-1970s. Slipper tanks were routine. (J. Roger Bentley Collection)

wore UK military serials. Two ex-airline Viscounts, a V.744 and a V.745D, flew for 10 years from 1962 with the Empire Test Pilot's School, based at Farnborough. Their highly demanding task was the practical training of ARB pilots in conducting aircraft certification tests. The Ministry of Technology purchased two used Viscounts – a V.837 and a V.838 – in 1964–65 for service with the Royal Radar Establishment at Pershore, near Worcester. Later, they passed to the Ministry of Defence, based at Bedford with the Royal Aircraft Establishment. Bedecked with various excrescencies, and believed primarily to have undertaken radar calibration work, the pair served until 1989 and 1991.

The Viscount as Test-Bed

For a time, the V.810 prototype, G-AOYV (c/n 316), provided the Vanguard program with an experimental vehicle. After first testing the control system, in 1959, the Viscount received a Vanguard tailplane section in place of its own vertical stabilizer. The tailplane portion was equipped with a Napier Spraymat deicing sys-

Still bearing residual United paintwork, despite reaching its fifth owner since leaving the airline, N7428 (c/n 126) was one of a number of U.S. Viscounts to find work with travel clubs such as Airworld. (BBA Collection)

Pearl Air, of the Bahamas, purchased c/n 248 in July 1975—when it was photographed at Maastricht, the Netherlands—registering the V.807 in Grenada as VQ-GAB. In Kenya, a different Pearl Air briefly owned a Zaire-registered Viscount. (Paul J. Hooper)

tem as fitted to the Vanguard. The test installation included a water spray rig mounted on the upper rear fuselage for deicing trials. The aircraft wore the Continental Airlines livery in which it had first served as a demonstrator, and, in due course, was reworked to airline standards and sold to VASP as a V.827.

United Aircraft of Canada – renamed Pratt & Whitney Aircraft of Canada – converted an ex-Air Canada V.757 (c/n 384) in 1972–73 to serve as a test-bed for the company's PT6A turboprop. The powerplant, complete with a modified DHC-7 nacelle, was installed in the Viscount's nose as a fifth engine, necessitating extensive strengthening of the aircraft's structure. For several years, the program concentrated on the PT6A-50 variant. After further nacelle adaptations, trials continued, now involving the smaller PT6A-28, –41 and –45 engine versions. The curiosity remained with P&WC until withdrawn in 1989.

Surprisingly, the interior of this Brazilian Air Force VC-90 seems a little less plush than one might expect, given that it was employed to transport the country's President. (Brooklands Museum)

Not all ex-BEA Viscounts survived to acquire full British Airways livery. At Cardiff in 1975, G-AOHS (c/n 167) is in the process of donating useful parts before going the way of its partner, believed to be G-AOJC (c/n 152). (BBA Collection)

British Airways retained a gradually dwindling Viscount fleet into the 1980s. G-AOYS (c/n 267) is seen in 1974 on climb-out; behind the inner propeller, note the small "Cambrian" titles—a vestige of pre-merger affiliation. (Global Air Image)

Many corporate Viscounts came finally to rest at Tucson, Arizona. N660RC (c/n 229), however, was still fully active when photographed there in April 1982, shortly after acquisition by Go Transportation. (Paul J. Hooper)

The RAAF employed two VIP-outfitted Viscounts during the 1960s; the second was this V.839 serialled A6-436 (c/n 436). Both aircraft were acquired second hand. (ATPH Transport Photos)

For a time, Dan-Air flew diverse Viscounts on a narrow range of schedules. V.807 G-CSZB (c/n 248), seen in mid 1979, was leased from Southern International. (Paul J. Hooper)

XR802 (c/n 198) of the ETPS is the object of some scrutiny at an open day. An ex-Capital V.745D, the aircraft possesses a radar nose, unlike its partner, XR801 (c/n 89). (BBA Collection)

A V.794D of the Turkish Air Force receives attention. 430 (c/n 430) served with THY as TC-SEL from 1958 until transferred to the military in 1971, and remained in operation in the mid-1990s. Interestingly, its serial is the same as the c/n. (BBA Collection)

At RAF Lossiemouth in February 1977, V.838 XT661 (c/n 371) of the Royal Radar Establishment wears external fittings very unlike those attached to its sister aircraft. Both aircraft carried slipper tanks; those of XT661 were of a noticeably more streamlined shape. (The apparent dorsal attachment is actually the fin of an HS Nimrod parked nearby.) (ATPH Transport Photos)

In the circuit at RAF Kinloss in May 1979, XT575 (c/n 438), one of two somewhat mysterious Viscounts that flew with the UK's Royal Radar Establishment, displays its large, pear-shaped radome. Markings included a red rear fuselage and fin, with a white rudder. (ATPH Transport Photos)

VICKERS
VISCOUNT

VISCOUNT VARIANTS

THE VICKERS TYPE NUMBERING SCHEME

Every Vickers design received a unique type number. Moreover, whenever modification of a design resulted in a new drawing, the amended variant earned a new number, no matter how slight the configuration differences, and even if it remained only a project. The system was purely chronological, and therefore intermingled versions of unconnected aircraft, which explains the occasional gaps in the progression of Viscount identities. The absent numbers between the V.708 and V.720, for instance, included variants of Valiant bomber.

In time, the system grew looser. "Round figures" (*e.g.*, V.800) came to denote a generic description, with the following batch of numbers indicating variations — hence V.801, V.802, etc., signified variants within the V.800 class. Specifically for the Viscount, a "D" suffix (*e.g.*, V.754D) was introduced to distinguish variants built to US certification standards.

Presentation of Viscount type numbers varies. Strictly, the "Viscount 814" is really the "(Vickers) Type 814." This book employs both styles, together with the generally accepted shorthand form with a "V." prefix, as in "V.814."

Type No.	Customer	Engines	Comments
453	-	Dart	original VC2 project deriving from civil Windsor; not built
609	-	4 x Mamba	VC2/Viceroy project; 2 prototypes; not built; redefined as V.630
630	MoS	Dart 502	prototype VC2/Viscount (G-AHRF); 2 airframes originally planned, one completed as V.663
640	Vickers	4 x Naiad	one prototype (G-AJZW); not completed
652	MoS	2 x Hercules	project; also proposed with 4 Hercules; not built
653	MoS	Dart 502	stretched-fuselage project; not built
663	MoS	2 x Tay	started as V.630; constructed as turbojet "Tay Viscount" (VX217)
700	MoS	Dart 504	one pre-production prototype (G-AMAV); built with major parts of uncompleted V.640; re-engined with Dart 505 and later with superior Dart versions
701	BEA	Dart 505	
701A	BEA	Dart 506	BEA designation for re-engined V.701; modified 1959-60 with airstairs and additional starboard window
701C	BEA	Dart 506	V.701 with Dart 506 installed during construction; later modified as per V.701A
700D	-	Dart 510	generic V.700 meeting US certification specs; term remained in use although officially superseded by V.770D
701X	BEA	Dart 506	BEA designation: 1 V.701A (G-AMOK) rebuilt/upgraded post-accident
702	BOAC Assoc. Co.	Dart 505	for BWIA; re-engined with Dart 506
703	BEA	Dart 505	project; high-density 53-seater with airstairs; not built
707	Aer Lingus	Dart 506	
708	Air France	Dart 505	
708A	Air France	Dart 506	re-engined V.708 (unofficial designation, *cf* V701A)
720	TAA	Dart 506	first Viscounts with increased fuel capacity; first 5 delivered with Dart 505 and later re-engined; cockpit windows converted to later design
721	Australian National A/W	Dart 506	optioned 1953; not built
723	Indian Air Force	Dart 506	part VIP interior
724	TCA	Dart 506	2 converted to executive V.744 by Timmins Aviation in 1963
728	BOAC Assoc. Co.	Dart 506	for Cyprus Airways, completed and painted but cancelled; then for Aden Airways but cancelled; altered to V.754D
730	Indian AF	Dart 506	full VIP interior
731	KLM	Dart 506	project; not built
732	Hunting-Clan	Dart 506	initially leased to MEA
734	Pakistan AF	Dart 506	VIP interior
735	Iraqi Airways	Dart 506	
736	Fred. Olsen	Dart 506	
737	Canadian DoT	Dart 506	
738	n/k	Dart 506	no further information; not built

Type	Customer	Engine	Notes
739	Misrair	Dart 506	
739A	Misrair	Dart 506	V.739 with minor upgrades including revised cockpit windows
739B	Misrair	Dart 506	V.739A with minor upgrades
740	UK Queen's Flight	Dart 506	project for staff aircraft; not built
741	UK Queen's Flight	Dart 506	project for VIP version of V.740; not built
742D	Braathens SAFE	Dart 510	completed and painted but cancelled; sold to FAB, full VIP cabin installed by Field Aircraft Services but type not re-designated
744	Capital	Dart 506	originally laid down as V.701
745	Capital	Dart 506	later re-engined with Dart 510 (*i.e.* converted to V.745D) by UAL
745D	Capital	Dart 510	final 15 not delivered but converted to various other V.770D for onward sale
746	East African	Dart 506	project; not built
747	Butler AT	Dart 506	
748D	CAA	Dart 510	DH propellers, 1st model delivered with weather radar installed
749	LAV	Dart 506	
750	-	Dart 520	project for high-speed, short-fuselage V.850 derivative; not built; type number originally officially allocated to a guided weapon project
754D	BOAC Assoc. Co.	Dart 510	for MEA; includes 2 originally completed as V.728; 1 other completed as V.794D
755D	Airwork	Dart 510	sold to Cubana pre-delivery
756D	TAA	Dart 510	
757	TCA	Dart 506	V.700D/V.770D standard except for powerplant
759D	Hunting-Clan	Dart 510	sold to Icelandair pre-delivery
760D	BOAC Assoc. Co.	Dart 510	for Hong Kong Airways
761D	UBA	Dart 510	
762D	-	Dart 510	executive version project; not built
763D	Hughes Tool Co.	Dart 510	executive requirement but airline interior; sold via intermediaries to TACA pre-delivery
764D	US Steel	Dart 510	executive use; auxiliary fuel tankage, airstairs
765D	Standard Oil	Dart 510	executive use; auxiliary fuel tankage, airstairs
766	Fred. Olsen	Dart 510	completed as V.779D
767	BOAC Assoc. Co.	?	project for Aden Airways; not built
768D	Indian Airlines	Dart 510	
769D	PLUNA	Dart 510	
770D	-	-	generic specification for North American model
771D	-	-	generic luxury/executive variant of V.770D
772	BOAC Assoc. Co.	Dart 506	for BWIA
773	Iraqi Airways	Dart 506	early cockpit window design
774D	-	Dart 510	project for Saudi Arabia; not built
775D	-	Dart 510	project for Yugoslavia; not built
776D	Kuwait Oil Co.	Dart 510	converted from completed V.745D
777	-	?	generic specification; no details known
778D	-	?	project for Chile; not built
779D	Fred. Olsen	Dart 510	not operated by owner; leased out
780D	-	Dart 510	generic specification for VIP version of V.700D
781D	South African AF	Dart 510	
782D	Iranian Airlines	Dart 510	included 1 with quick-change airline/VIP cabin
783D	LAV	Dart 510	project; not built
784D	PAL	Dart 510	includes 1 converted from completed V.745D after being leased out
785D	LAI	Dart 510	all converted from completed V.745D; taken over by Alitalia
786D	LAC	Dart 510	sold to LANICA and TACA pre-delivery
787D	Iraqi Airways	Dart 510	project; not built
788D	Syrian Airways	Dart 510	project; not built
789D	FAB	Dart 510	part-VIP interior
790	-	Dart 506	generic specification for Local Service Viscount project; not developed; c/n 420 allocated for prototype but not re-used
791	AVIANCA	Dart 506	project; not built
792	Pakistan AF	Dart 506	project; not built
793D	Royal Bank of Canada	Dart 510	Originally V.745D and leased out, then converted for RBC
794D	BOAC Assoc. Co.	Dart 510	for THY; 1 converted from V.754D purchased direct
795D	TWA	Dart 510	projected conversion of completed V.745Ds; not proceeded with
796D	Turkish AF	Dart 510	VIP project; not built
797D	Canadian DoT	Dart 510	converted from completed V.745D; VIP interior
798D	Northeast	Dart 510	includes 8 aircraft converted from completed V.745Ds
800	-	Dart 510	generic initial specification for stretched Viscounts; after dimensions revised to type 877 specification, V.800 continued to be used as basic term embracing V.802-809
801	BEA	Dart 508?	project to original type 800 specifications; not built

802	BEA	Dart 510	definitive fuselage stretch to V.877 specifications
803	KLM	Dart 510	
804	Transair	Dart 510	
805	Eagle Airways	Dart 510	converted to V.808 at later purchase by Aer Lingus
806	BEA	Dart 520	nine later re-engined with Dart 510
806A	BEA	Dart 520	1 aircraft only (G-AOYF), loaned to Vickers as prototype for V.806 and development for V.810; destroyed; airframe rebuilt as a new V.806
807	NZNAC	Dart 510	
807B	NZNAC	Dart 510	1 aircraft converted from V.804 to V.807
808	Aer Lingus	Dart 510	
808C	Aer Lingus	Dart 510	V.808 converted by Scottish Aviation to cargo configuration
809	-	Dart 510	project for Greece; not built
810	-	Dart 525	generic specification; type number used for 1 prototype (G-AOYV) with rear lounge, later converted to V.827 for sale
810D	-	Dart 525	D suffix briefly used to denote "Americanized" V.810s; discontinued
810U	-	Dart 525	allocated temporarily to V.810s built 1959-60 without official orders
811	Capital	Dart 525	project; not built
812	Continental	Dart 525	first with lounge in place of rear baggage hold; airstairs; 11 aircraft later fitted with Dart 530 by Channel Airways
813	SAA	Dart 525	no lounge; re-engined with Dart 530
814	Lufthansa	Dart 525	no lounge
815	PIA	Dart 525	no lounge; 2 re-engined with Dart 530 before onward sale
816	TAA	Dart 525	rear lounge; airstairs; 3 undelivered aircraft converted — 1 to V.836, and 2 to V.839 — and sold
817	AVENSA	Dart 525	project; not built
818	Cubana	Dart 525	1 aircraft converted to V.835 and sold pre-delivery
819	Niarchos Group	Dart 525	executive variant; order cancelled before construction; c/n reallocated to a V.814
820	American Airlines	Dart 525	project; not built
821	Eagle Aviation	Dart 525	order cancelled before construction
822	LAV	Dart 525	project; not built
823	California Eastern	Dart 525	order cancelled before construction
824	LAI	Dart 525	project; LAI taken over by Alitalia; aircraft not built
825	Black Lion Aviation	Dart 525	order cancelled before construction although some sub-assemblies manufactured
826	Aigle Azur	Dart 525	project; not built
827	VASP	Dart 525	rear lounge; includes 1 converted from V.810 prototype
828	All Nippon	Dart 525	rear lounge; TV installed (type number originally allocated to Rolls-Royce for test-bed aircraft; construction started but order lapsed)
829	TAP	Dart 525	order unconfirmed; not built
830	American Airlines	Dart 525	project with extended fuselage; not built
831	Airwork	Dart 525	1 supplied direct to Sudan on lease; re-engined with Dart 530; no lounge
832	Ansett-ANA	Dart 525	rear lounge; 2 aircraft converted to V.837 during construction
833	Hunting-Clan	Dart 530	no lounge; single window ahead of starboard props
834	LOT	Dart 525	project; not built
835	Tennessee Gas	Dart 525	converted from completed V.818; VIP interior
836	Union Carbide	Dart 525	converted from completed V.816; VIP interior
837	AUA	Dart 525	originally V.810U except for 2 converted from V.832 during construction; rear lounge
838	Ghana Airways	Dart 525	originally V.810U; no lounge
839	Iran Govt/Union Carbide	Dart 525	1 each operator; converted from V.816 during construction; rear lounge; VIP interior (type number allocated originally to All Nippon)
840	-	Dart 541	generic specification for more powerful version of V.810
841	BEA	Dart 541	option placed but dropped; not built
842	Iraqi Airways	Dart 541	option placed but dropped; not built
843	CAAC	Dart 525	variant of V.810; rear lounge
844	Liberian Govt.	Dart 541	option placed but dropped; not built
845	PIA	?	proposal; not built (variant of V.810 or V.840?)
850	-	Dart 550	further-stretched fuselage (Viscount Major); basic project only
860	-	Dart 525	study for V.850 with earlier engines
870	-	Tyne	studies for larger, heavier Viscount developed from V.850, also including turbojet-powered variant; led to V.900 and V.950 (*i.e.* Vanguard)
877	-	Dart 510	denotes revised V.800 specification as actually built, with 3ft. 10in. fuselage extension over V.700

SPECIFICATIONS

PRINCIPAL VISCOUNT AND ROLLS-ROYCE DART DATA

	V.630	V.700	V.700D	V.800	V.810
DIMENSIONS:					
Length without radar	74 ft 6in	81 ft 2 in	—	85 ft 0 in	—
with radar	—	81 ft 10 in	81 ft 10 in	85 ft 6 in	85 ft 8 in
Wingspan	89 ft 0 in	93 ft 8.5 in	93 ft 8.5 in	93 ft 8.5 in	93 ft 8.5 in
Tailplane span	37 ft 0 in	37 ft 0 in	37 ft 0 in	37 ft 0 in	37 ft 0 in
Height to top of fin	26 ft 3 in	26 ft 9 in	26 ft 9 in	26 ft 9 in	26 ft 9 in
Wing area	885 sq ft	963 sq ft	963 sq ft	963 sq ft	963 sq ft
Tailplane & elevator area	—	238 sq ft	238 sq ft	238 sq ft	238 sq ft
Fin & rudder area	—	124 sq ft	124 sq ft	124 sq ft	158 sq ft
Fuselage internal diameter	10 ft 0 in	10 ft 0 in	10 ft 0 in	10 ft 0 in	10 ft 0 in
Cabin length standard	—	45 ft 0 in	45 ft 0 in	54 ft 0 in	54 ft 0 in
including rear lounge	—	—	—	—	56 ft 6 in
Cabin width	9 ft 7 in	9 ft 7 in	9 ft 7 in	9 ft 11 in	9 ft 11 in
Underfloor hold volume	—	215 cu ft	215 cu ft	250 cu ft	250 cu ft
Rear hold volume	—	155 cu ft	155 cu ft	120 cu ft	120 cu ft
Forward door (inches)	—	64 X 48	64 X 48	64 X 60	64 X 36
Rear door (inches)	—	66 X 54	66 X 54	64 X 27	64 X 27
Powerplant	R.Da.1	R.Da.3	R.Da.6	R.Da.6	R.Da.7/1
	Mk 502	Mk. 505/506	Mk. 510	Mk. 510	Mk. 525/530
Max. takeoff weight (lb)	45,000	63,000	64,500	64,500	72,500
Max. landing weight (lb)	40,000	58,500	58,500	58,500	62,000
Max. zero fuel weight (lb)	36,000	49,000	50,168	55,000	57,500
Empty weight (basic equipped)	27,000	36,860	37,920	41,200	43,200
Max payload (lb)	9,000	12,140	12,250	13,700	15,000
Passenger seating (at delivery)	32	40–53	40–53	53–65	56–64
PERFORMANCE:					
Max. payload range					
w/ reserves (miles)	700	970	1,330	690	1,275
w/ slipper tanks (miles)	—	1,110	1,470	—	1,750
Max. fuel range					
w/ reserves (miles)	1,100	1,200	1,720	1,340	1,350
w/slipper tanks (miles)	—	1,450	1,870	—	1,750
w/ slipper & belly tanks (miles)	—	—	2,450	—	—
Max cruise speed (mph)	300	318 (Dart 505)	334	325	365
		324 (Dart 506)			
Economical cruise speed (mph)	273	302 (Dart 505)	324	310	351
		316 (Dart 506)			
Service ceiling (ft)	—	28,500	27,500	27,000	27,000
Field length (takeoff to 35 ft)					
at ISA w/o water-methanol) (ft)	4,600	5,500	5,600	5,450	6,100
Takeoff speed at MTOW (mph)	112	132	131	133	136
Rate of climb at MTOW (ft/min)	1,100	1,200	1,400	1,220	1,650
Approach speed at MLW (mph)	114	133	132	134	137
Landing run (from 50 ft at ISA) (ft)	4,000	4,200	4,200	4,200	4,300

The Rolls-Royce Dart propeller turbine was categorized by series (also called the rating designation) and mark. Customarily, the mark number alone was used to identify individual types. The Dart variants employed in the Vickers Viscount were as follows:

Series	Mark	Turbine Stages	Reduction Gear Ratio	Takeoff EHP	Continuous EHP	Cruise EHP	Viscount Type Application
R.Da.1	502	2	0.091	1,380	—	—	V.630
R.Da.3	504	2	0.106	1,547	—	—	V.700 prototype
R.Da.3	505	2	0.106	1,547	1,190	890	V.700
R.Da.3	506	2	0.106	1,547	1,190	955	V.700
R.Da.6	510	2	0.093	1,740	1,435	1,025	V.700D/V.800
R.Da.7	520	3	0.093	1,890	1,655	1,145	V.806
R.Da.7	525	3	0.093	1,990	1,745	1,325	V.810
R.Da.7	530	3	0.093	2,030	1,745	1,325	V.810

VICKERS VISCOUNT

MODEL KIT GUIDE

BY RICHARD MARMO

Some aircraft that hold an important place in aviation history, such as the Vickers Viscount, have gotten short shrift as subjects of plastic model kits. At one extreme is the Boeing 707, which has been produced in model form by more manufacturers than you can count. At the opposite end of the scale is the Viscount.

As far as can be determined, the only widely available styrene kit of the Viscount was in 1/96th scale and originally produced by Hawk. It wound up in eight different boxes under the Hawk label, the main differences being box art, price (remember, manufacturers used to print retail prices on the boxes), and decals (Capital Airlines, UAL and Continental). This kit was available from Hawk up until 1971.

According to *The Collectors Value Guide For Scale Model Plastic Kits* by John W. Burns, during the 1950s, this same Viscount kit was available two other ways. First, in a mail order box from Capital Hobbies with Capital Airlines markings. Secondly, as a promotional item in a mailer box directly from Northeast Airlines that featured – obviously – Northeast Airlines markings on the decal sheet. They're both rare (I've never seen one) and pricey, with values into three figures.

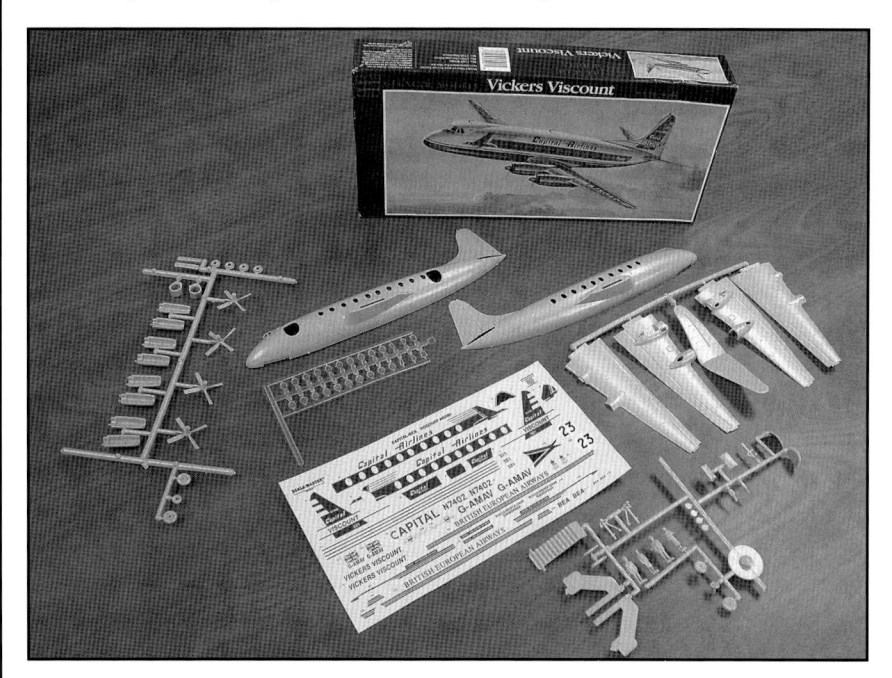

The only styrene kit of the Vickers Viscount was a Hawk kit. The molds were later acquired by Glencoe Models, which produces the kit with this box art and decals. The molds have held up well, and the fit and detail is quite good. Original Hawk kits are rare and expensive, but the Glencoe versions are still widely available and affordable.

When Testors bought Hawk in 1971, the Viscount was one of those Hawk kits that wasn't reissued, and most airline enthusiasts thought the kit was gone for good. Fortunately, it surfaced again in the late '80s as a Glencoe Models kit. This time around, the kit included Scale-Master decals for Capital Airlines and British European Airways versions. At the same time, Glencoe Models issued three aftermarket decal sheets produced, as you'd expect, by Scale-Master, and all designed to fit the Viscount kit. Two sheets were released in 1989 and the third in 1990.

The first sheet, #05501-1, provided markings for Northeast Airlines and Air France. Sheet #05501-2 was somewhat larger, with markings for three different airlines: Aer Lingus, Alitalia, and Trans Australia Airlines. Finally, #05501-3, the smallest of the group, allowed you to choose between Air Canada and Trans-Canada Air Lines.

When it comes to finding one of these kits to add to your collection, the effort ranges all the way from difficult and expensive to easy and economical, depending on what you're after. If you're chasing an original Hawk kit, it's likely that you'll be dealing with Ebay or a private collector and writing a check for $60 to $100 in the process.

Luckily, the Glencoe version of the kit is still in production and much easier to find, and at reasonable prices. Five minutes on an Internet search engine will net you several mail order companies that list the Viscount and most of the other Glencoe kits at prices very close to normal retail. Prefer Ebay? You'll find Glencoe Viscounts there for as little as $8 to $10, and at that price you can actually afford to *build* the kit.

Although the Viscount has only been replicated in a single injected styrene plastic kit, that doesn't mean there aren't other models of it. One other kit does exist – a *paper* model from a German company, J.F. Schreiber. It's 1/100th scale and sells for $19.95. I haven't seen this particular kit, but if it's anything like other paper aircraft models I've seen, the result should be quite realistic. Just be aware that you'll have to develop a new set of modeling skills.

Finally, Corgi offers a series of 1/144th-scale Viscount replicas that are aimed at the collector. Keep in mind that these are not kits but collectibles. The intent is that you buy the model, take it out of the box, set it in a display case, and admire it. No building involved. At the present time, I believe that these die cast Viscounts are available in either six or eight different liveries.

If you want to learn some new modeling skills, try this paper Vickers Viscount kit in Austrian Airlines livery from J.F. Schreiber. This German company has been making paper card kits of all kinds of subjects for more than 100 years.

There have been a number of other Viscount kits/models produced over the years. They're rare at best, are generally of interest only to the serious collector, and can be very expensive. A 1/58th scale tinplate toy was produced by Tomiyama, Airways chimed in with a 1/72nd vacuform, and manufacturers such as Co-Ma/Aermac, J&L, Lincoln, Faller, Frog, Sebel, Welsh, and others released their own efforts. Scales were all over the landscape, including 1/100th, 1/121st, and 1/144th. Media runs from crude vacuforms thru resin to semi-finished display models with no landing gear.

Given the rate at which limited edition kits of some truly exotic (and virtually unknown) aircraft types are being announced of late, I wouldn't be at all surprised if a new Vickers Viscount kit (either plastic or resin) eventually appears in a popular scale.

Sources

Corgi Classics, Inc.
175 W. Jackson Blvd., Suite 1770
Chicago, IL 60604
312-873-0348
www.corgiclassics.com

Glencoe Models
Box 846
Northboro, MA 01532
508-869-6877
www.glencoemodels.com

J.F. Schreiber
Schreiber-Bogen
Aue-Verlag
Schottstr. 52
D-70192 Stuttgart
Germany
www.schreiber-bogen.de

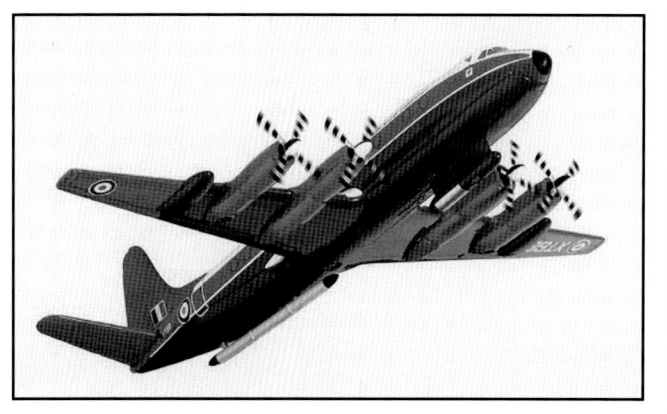

Corgi makes a nice line of display-ready, diecast aviation models, including several versions of the Viscount.

SIGNIFICANT DATES

KEY DATES IN THE HISTORY OF THE VICKERS VISCOUNT

December 1942
The Brabazon Committee defines airliner requirements for the UK's postwar needs.

13 March 1945
Lord Brabazon recommends Type 2B, a 24-seat aircraft, with four gas turbine engines, for European and other short-to-medium range services.

19 April 1945
MAP instructs Vickers to develop the VC2 prototypes – pressurized and unpressurized 24-seaters and an unpressurized 27-seater.

20 September 1945
A Gloster Meteor, powered by two Rolls-Royce Trent propeller-turbines, makes the world's first turboprop flight.

17 April 1946
The MoS issues Specification 8/46 to Vickers for a Brabazon Type 2B airliner: the powerplant is the Armstrong Siddeley Mamba.

December 1946
Construction of the type 609 – the VC2 Viceroy – begins.

27 August 1947
The MoS alters the specified VC2 powerplant to the Rolls-Royce Dart.

10 October 1947
The Rolls-Royce Dart turboprop flies for the first time, in the nose of an Avro Lancaster.

December 1947
British European Airways orders the rival Airspeed Ambassador; Vickers counters with revised proposals. The government rejects them and the Viscount is sidelined.

16 July 1948
At Wisley, the V.630, registered G-AHRF, makes its first flight.

2 September 1948
V.630 gives its first public demonstrations, and a week later flies at the SBAC air show at Farnborough.

January 1949
Enlarged and improved Viscount, the V.700, is announced; MoS orders one prototype.

August 1949
Viscount 630 is demonstrated to BEA, whose enthusiasm has been revived by the V.700.

15 September 1949
The V.630 is awarded the first Certificate of Airworthiness issued to a turbine-powered air transport.

15 March 1950
The second Viscount – the V.663 "Tay Viscount," powered by two Rolls-Royce Tay turbojets – makes its first flight.

20 March 1950
The V.630 begins a demonstration tour of typical BEA European destinations.

29 July 1950
V.630 G-AHRF operates the world's first turboprop scheduled airline service, flying for BEA from London (Northolt) to Paris (Le Bourget).

3 August 1950
BEA places the first Viscount order, contracting for 20 V.701s.

28 August 1950
The prototype of the definitive V.700, G-AMAV, makes its first flight.

15 August 1951
In-service powerplant trials begin, using two Dart-engined C-47s, operating scheduled cargo services for BEA.

3 October 1951
The prototype V.700 is granted a C of A.

20 August 1952
The first production Viscount, V.701 G-ALWE, makes its initial flight.

27 August 1952
The V.630 is damaged beyond repair in a landing accident during trials at Khartoum.

November 1952
Trans-Canada Air Lines orders 15 of a reworked Viscount version, the V.724.

20 January 1953
BEA formally accepts the first turboprop airliner when it receives G-ALWE.

13 February 1953
Bound for trials in Canada, G-AMAV conducts the first trans-Atlantic flight by a turboprop aircraft.

18 April 1953
The world's first sustained turboprop scheduled services begin when the V.701 G-AMNY departs on a BEA service from London to Rome, Athens, and Nicosia.

8–10 October 1953
G-AMAV participates in the UK-New Zealand air race on behalf of BEA.

1 December 1953
The first Viscount built at the new factory at Hurn – G-AMOO – takes off for the first time.

14 April 1954
BEA orders 12 V.802s, a stretched variant with a fuselage extension of slightly less than four feet.

11 August 1954
Capital Airlines confirms an order for 37 Viscount 745Ds, having first contracted for three V.744s in May.

18 December 1954
Opening scheduled Viscount services, TAA becomes the first non-European airline to sell fares on a turbine airliner.

4 April 1955
A TCA V.724 operates the first turbine-powered scheduled commercial air service in North America.

13 June 1955
The Viscount 700D receives its US Type Certificate from the Civil Aeronautics Administration.

26 July 1955
US turbine-powered domestic scheduled services commence, when a Capital Airlines Viscount 744 leaves Washington, DC, for Chicago.

27 July 1956
The first stretched Viscount – a V.802 for BEA, registered G-AOJA – makes its first flight.

29 November 1956
The first executive-configured Viscount – N906, a V.764D for the US Steel Corp. – is delivered.

18 February 1957
A London to Paris flight by a BEA V.802, G-AOJE, marks the first passenger service by a long-fuselage Viscount.

14 March 1957
The first production Viscount, G-ALWE of BEA, crashes on approach to Manchester (Ringway).

23 December 1957
G-AOYV, the first Viscount 810, featuring a stronger structure, Dart R.Da.7/1 Mk 525 engines, and a 365 mph cruising speed, makes its first flight.

27 May 1958
Granted a US Type Certificate on 22 April, the top-of-the-range Viscount 810 enters service with Continental Airlines.

14 May 1959
G-AOYS, a V.806 of BEA, operates the first scheduled airline service between the UK and the USSR, connecting London and Moscow via Copenhagen.

29 May 1959
The 400th Viscount – G-APTB, the first V.833 for Hunting-Clan – is delivered.

16 April 1964
B-412 (as G-ASDV) of CAAC, constructed at Hurn, is the 444th and final Viscount delivered.

18 April 1996
On the 43rd anniversary of the opening of sustained Viscount services, G-APEY – a V.806 of British World Airways first delivered to BEA – operates the UK's final Viscount passenger flight.